ATLAS OF THE SHEEP BRAIN

Second Edition

Text by R. Glenn Northcutt
Dissection by Kenneth L. Williams
Photography by Robert P. Barber

ISBN 978-1-58874-695-5

Special Printing 2013

Published by
Stipes
PUBLISHING
Champaign, Illinois

DEDICATED TO THE MEMORY OF

JOHN WILLIAM SHEHORN

who opened the door for a
young man to the beauty of
nature and a faith in man,
but could not follow.

PREFACE

With the recent increase of interest in neurology, there is ever growing need for a manual which goes beyond the brief presentation given in beginning anatomy texts, but is less extensive than the average text in neurology. It is hoped that this atlas will to some extent meet this need and be of use to students in many fields who need a basic understanding of the rudiments of neuroanatomy.

This manual has been designed with the assumption that the student is at least acquainted with basic anatomical terminology that he might be expected to have obtained at the end of a general anatomy course. It is in no way meant to replace the many extensive texts on neurology but is to be used as an aid to bridge the gap between lecture and laboratory work, and to summarize the functional complexity of the brain.

In studying the anatomy and function of the brain, the student is faced with two approaches. He may study the regional anatomy of the brain, which leaves him with the problem of fitting together functions at his leisure. On the other hand, he may study the brain by functional systems, tracing all structures involved in a common function. However, the latter method leaves him with the equally difficult task of fitting together structural relationships. It is the opinion of the author that the second method has the fewest disadvantages. By tracing functional systems, a certain amount of repetition is inevitable. This repetition, however, can serve to firmly commit the general structures to memory while also giving the student a general idea of their functional significance, so important to any understanding of the evolution and function of the nervous system.

As in any work, the author is deeply indebted to many individuals for their encouragement and interest. I wish to thank Drs. Clyde A. Dennis and Lester C. Shell of the Department of Biology, Millikin University, who kindly made facilities available for this work. I am equally indebted to Mr. William P. Anderson who greatly enriched my teaching experience while at Millikin, where many of the methods of approach presented in this atlas came to mind. Finally I wish to express my gratitude to Dr. Garth J. Thomas of the Department of Physiology and Biophysics, University of Illinois, and Mr. Bruce B. Criley who critically read the manuscript and offered many valuable corrections and suggestions. The author while gratefully recognizing the assistance of many, nevertheless accepts full responsibility for any errors or omissions in the final result.

Urbana, Illinois
July, 1964

R. G. N.

TABLE OF CONTENTS

I. MATERIALS AND METHODS

The sheep brain was chosen for presentation because it can be obtained cheaply in large numbers and is large enough to be dissected easily. This is not the case with cat brains which are difficult to study macroscopically or with monkey brains which can not be obtained in large numbers.

This atlas may be used in conjunction with several methods of study. If time permits, each student can be given several brains. One brain can be dissected freely with little or no directions given. This gives the student a general feeling for structural relationships. This introductory phase can be followed by carefully guided dissection following the figures as they occur in the atlas.

If the time allowed for study of the nervous system is short, a single brain can be used to identify all listed structures in the manual. The student should first examine all external structures taking care that he not only memorizes the terms listed but that he gains some idea of their interrelationships and functional significance. Once the external surface has been studied, the brain should be divided into left and right halves using a brain knife or some other suitable instrument and cutting midsagittally through the brain with a smooth clean stroke. A lubricant such as glycerine can be smeared on the knife if resistance is encountered while cutting.

The right half of the brain should be cut parasagittally following the directions given in the insert of Figure 8. The two pieces of the right side of the brain should then be held together, turned with the lateral surface uppermost and the horizontal sections cut following the directions given in the inserts of Figures 9 and 10. If the above directions have been followed correctly, the right half of the brain will be divided into six pieces which can be reconstructed to study the above mentioned figures.

The left half of the brain should then be cut transversely following the directions given in Figure 11 which represents the positions of the remaining figures of the atlas. Throughout the study, care must be taken that the brains do not dry out as they become useless when so allowed.

All structures denoted in this atlas can be observed without macroscopic staining but it is recommended that the staining be undertaken as it increases the contrast of the structures mentioned.

The LeMasurier modification of the Mulligan method, which stains the gray matter blue while leaving the white matter unstained, was used because of the rapidity and simplicity of the method. It was found that students with little or no experience in staining technique have few problems in obtaining excellent results with this method.

The brain sections after being cut are placed in running water overnight to remove the preservative and then placed in three changes of distilled water for one hour. The sections are then placed in preheated Mulligan's solution at 60°C. for two to five minutes. This solution forms a protective film over the white matter, so that it will not take up the stain. The sections are then washed in a large volume of cold tap water for one minute and placed in a 1 per cent solution of ferric chloride in distilled water for two minutes. Then they are removed to running tap water for five minutes and placed in a 1 per cent solution of potassium ferrocyanide in distilled water until the gray matter turns a brilliant blue. This should take approximately three minutes. The sections are then washed in running water for twenty-four hours and preserved in 70 per cent alcohol. Sections so preserved will retain their color for years and can be stored and used for demonstrations or practical examinations.

Solutions

Mulligan's solution

Carbolic acid (phenol), crystalline	40.00 gm.
Copper sulphate, crystalline	5.00 gm.
Hydrochloric acid, conc.	1.25 cc.
Water, distilled	1000.00 cc.

Do not breathe fumes for long periods as they are toxic.

Ferric chloride (1% solution)

Ferric chloride, crystalline	10.00 gm.
Water, distilled	1000.00 cc.

This solution should be made up just before use as it decomposes rapidly.

Potassium ferrocyanide (1% solution)

Potassium ferrocyanide, crystalline	10.00 gm.
Water, distilled	1000.00 cc.

This solution is stable and can be stored.

II. NEURAL MECHANISMS

To attempt in a few pages to describe the functional complexities of the nervous system is a difficult task. However some attempt must be made to draw anatomical and physiological aspects of the central nervous system into an integrated whole. This approach will give the beginning student some idea of the organization of the brain and its role in the integration of other organ systems.

This discussion will depend to a great extent on the diagrams which accompany it. Such a presentation has both its strong and weak points. It is useful in that diagrams give the student a compact and clear idea of the functional connections found within the system. It fails or at least is a burden to the student in that once mastered the student assumes he understands the nervous system in all its complexity. This often results in the student carrying more erroneous concepts away from the study than he brought to it. If this shortcoming of the diagrammatic method is kept in mind and the student can remember to throw away this diagrammatic crutch when he no longer needs it, this method has served a useful purpose in the course of his progress to an understanding of this system.

The use of the term neural mechanism should be discussed and some of its implications noted before further discussion of the nervous system is undertaken. This term is used to delimit the functional properties of an arbitrarily limited anatomical structure or structures. An example of a neural mechanism is the coordination of muscular movement in phasic balance as mediated by the cerebellum. Only specific areas of the cerebellum are involved in this function, not the entire cerebellum. Those parts of the cerebellum and its connections that are concerned with this function can be discussed under a single heading. Such an anatomical and physiological complex is a neural mechanism. Two weaknesses of this arbitrary method of approach must be kept in mind.

First of all, an understanding of the functional role of a structure is no better than the student's understanding of the structure on which the function rests. Secondly the division into mechanisms is an artificial one. The brain functions as a unit. Every functional mechanism is reciprocally affected by its counterparts, i. e., no single mechanism is independent or unaffected by other mechanisms.

It should be stressed that a number of terms and concepts must be understood in order to follow the discussion and to understand the conventions used in the diagrams. The basic anatomical and physiological unit of the nervous system is the neuron, which consists of a cell body or perikaryon and its processes. There are two types of processes, dendrites and axons. The dendrites are usually short in length and highly branched, whereas the single axon is long and relatively unbranched except at its termination. The dendrites and cell body will be represented by a circle in all diagrams in this text. The axon will be represented by a heavy line drawn from one circle (cell body and dendrites) to another circle, representing another cell body or organ. The function of the neuron consists of integration of presynaptic impulses resulting in transmission of an impulse which is always conducted from the dendrite or cell body along the axon away from the cell body.

Neurons are not randomly scattered throughout the nervous system but are arranged in groups having a similar structure and function. Such a group of neurons within the brain or spinal cord is called a nucleus. A similar group of neurons found outside the brain or spinal cord is called a ganglion. A group of axons which connect neurons of one nucleus with the neurons of another nucleus within the brain or spinal cord is called a tract. A similar group of axons connecting one ganglion with another ganglion or effector organ is called a nerve.

In mammals, most axons are surrounded by supporting cells, called glial cells which lay down a myelin sheath and give the tracts a white appearance in contrast to the nonmyelinated cell bodies which have a gray appearance. This results in gray and white areas within the nervous system. The gray areas primarily consist of cell bodies of neurons, dendrites, and nonmyelinated axons; the white areas primarily of myelinated axons.

The nuclei, represented as stippled areas in the diagrams, are named after their discoverers or by a descriptive term. The tracts are represented by one or more lines connecting one nucleus to another. It should be remembered that a single nucleus and its tracts are composed of thousands of neurons and axons, not just the few represented in the diagrams. The tracts are sometimes descriptive in their nomenclature. Many tracts have compound names derived from their nucleus of origin and nucleus of termination. An example of the latter type is the corticopontine tract. This tract begins in the cortex of the cerebrum and ends in the pontine nuclei of the pons.

In addition several descriptive terms are used in reference to both tracts and nuclei. The terms sensory, associative (integrative), and motor are commonly used. A sensory nucleus or tract receives and transmits coded sensory impulses (external or internal stimuli) from a sensory end organ to integrative nuclei and tracts as well as to motor nuclei and tracts. Integrative nuclei receive impulses from sensory and other integrative nuclei and often modify these impulses before transmitting them to other integrative nuclei or motor nuclei. The motor nuclei transmit impulses to muscles and glands which result in contraction and secretion of these same organs.

The terms afferent or ascending and efferent or descending are often used to describe the projection of tracts to or from nuclei in the brain and spinal cord. A tract projecting to the brain or to a nucleus is said to be ascending to the brain or afferent to a nucleus. A tract projecting away from the brain or a nucleus is said to be descending from the brain or efferent from a nucleus.

SPINAL MECHANISMS

A knowledge of the spinal connective patterns are of great importance to any understanding of the functions of the brain. These simple connective patterns are the final common patterns used by all higher pathways, both afferent and efferent, to bring their influence upon all organs of the body. Sherrington's reflex concepts were primarily responsible for an understanding of these innate functional patterns. In a spinal animal, an animal in which the connections between brain and spinal cord have been severed, these reflex patterns are isolated from the influence of higher centers and are thus easily studied. At present, they give us the clearest picture of the functional connections or "wiring" of the spinal cord but also lead to several misconceptions. For example, observation of these reflexes give the impression that they are stereotyped or inflexible, which is completely erroneous. One of the most plausible generalizations that can be made regarding the nervous system is that it is a highly flexible system. These patterns of movement or reflexes call our attention to the major primary functions of the spinal cord but give no idea of the extreme flexibility that exists in spinal mechanisms. Secondly, it is easy to assume erroneously that functions of the brain are merely made up of numerous multiple reflex patterns. If such a view is taken the student ends up with a model of stereotyped patterns resulting in an organism rigidly acting like a machine. Such a view does not account for the extreme plastic behavior of higher forms or the adaptability of the nervous system. However, the reflex concept does serve two primary uses. It is useful to the clinician in the analysis of pathological conditions and to the neurologist as a tool for obtaining some idea of the major connections and their resultant interactions.

These reflexes call our attention to two basic types of patterns: the monosynaptic arc and the multisynaptic arc. The monosynaptic arc, although the latest to evolve in vertebrates, will be considered first due to its simplicity. The monosynaptic arc (Diagram 1) or myotatic reflex consists of a sensory neuron located in the dorsal root ganglion and a motor neuron located in the ventral gray column of the spinal cord. The dendritic endings of these sensory neurons are modified to monitor changes in the stretch of muscles or changes in cutaneous sensation. The modified dendritic endings found in muscle are called annulospiral endings. Those found in the skin fall into several categories which will be discussed further under somatic sensory mechanisms. The sensory neurons of the myotatic reflex pass their axons into the gray columns of the spinal cord where they form synaptic connections with large motor neurons which are called alpha motor neurons. The synaptic junctions formed fall into two categories. The alpha motor neurons may be facilitated (represented in Diagram 1 by a plus sign) to transmit an impulse or they may be inhibited (designated by a minus sign) from transmitting. A single sensory neuron alone usually can not cause a single motor neuron to transmit an impulse to a muscle. Hundreds of synaptic endings acting on a pool of motor neurons are necessary to result in a "train" of firing which in turn gives rise to a smooth contraction of muscle tissue. In discussing such reflex patterns, the convention is usually adopted of referring only to a single sensory or motor neuron. However, it is essential to remember that the neuron referred to is but a single member of a vast population of such neurons involved in a common functional pool.

Those alpha neurons which innervate extensor muscles are facilitated by sensory stimuli from annulospiral endings, whereas those neurons which innervate flexor muscles are inhibited by sensory stimuli from annulospiral endings. Therefore contraction of an extensor muscle coincides with relaxation of a corresponding flexor muscle. This functional condition of the spinal cord is referred to as reciprocal innervation and reflects a basic condition found throughout the nervous system of vertebrates.

The myotatic reflex is the spinal mechanism responsible for reflex tone and posture in the tetrapodal vertebrates. The stretch on extensor muscles by gravity results in stimulation of the annulospiral endings which stimulate alpha motor neurons to fire, causing contraction of the extensor muscles in the limbs.

The alpha motor neurons do not fire continuously, but are modulated by their sensory input plus neuronal patterns which form a negative feedback system. This system is formed by the Renshaw cells. These cells, found in the ventral gray column, receive impulses from the alpha motor neurons via axonic collaterals which branch from the axon and form secondary connections. When an alpha motor neuron fires, Renshaw cells fire back on the alpha motor neuron thereby temporarily inhibiting additional firings of the alpha motor neuron.

Finally, one other factor can affect the myotatic reflex by way of its sensory input. The information transmitted regarding the amount of stretch to the alpha neuron can be varied by independent contraction of the few muscle fibers innervated by annulospiral endings in a given muscle. This contraction is brought about by stimulation of gamma motor neurons in the ventral gray column of the cord (not represented in Diagram 1). Their afferent input primarily originates from higher centers and will be discussed later under reticular mechanisms.

Diagram 1. Spinal Patterns

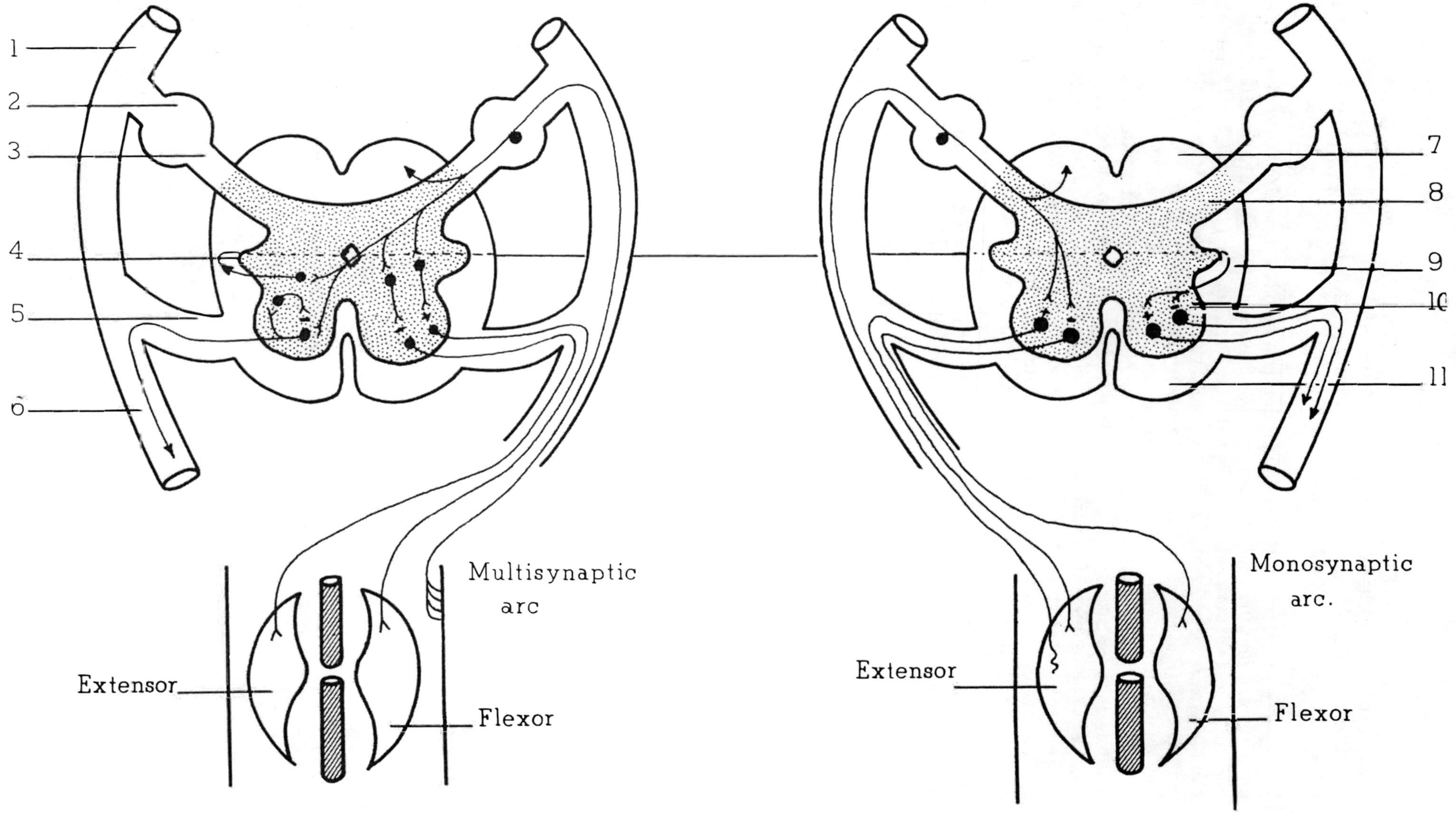

DIAGRAM 1: SPINAL PATTERNS

1. Dorsal somatic branch of spinal nerve
2. Dorsal root ganglion
3. Dorsal root of spinal nerve
4. Lateral gray column
5. Ventral root of spinal nerve
6. Ventral somatic branch of spinal nerve
7. Dorsal funiculus
8. Dorsal gray column
9. Lateral funiculus
10. Ventral gray column
11. Ventral funiculus

The remaining spinal reflex types are based on the multisynaptic pattern. Each of these reflexes has a similar neuronal pattern which is based on a sensory neuron, one or more associative or internuncial neurons, and a motor neuron. The flexion reflex is characteristic of this neuronal pattern and will be considered first. Its sensory neuron is located in the dorsal root ganglion. Its dendritic endings may be cutaneous or deep in connective tissue. Transmission is elicited by noxious stimuli such as burning, pinching or other chemical and physical stimuli. The sensory neuron passes its axon into the dorsal gray column where it synapses with one or more internuncial neurons. These internuncial neurons facilitate alpha motor neurons to flexor muscles and inhibit alpha motor neurons to extensor muscles. The response results in contraction of flexor muscles so that the limb stimulated by the noxious stimulus is withdrawn. Such a reflex obviously has protective value. Contraction of the limb on the same side of the animal as the noxious stimulus occurred is called an ipsilateral flexion reflex.

This is contrasted with a contralateral reflex in which the internuncial neuron may pass across the dorsal gray commissure and synapse with motor neurons of the opposite ventral gray column. In addition, the internuncial neurons may pass from one segment where the noxious stimulus occurs and result in contraction of flexor muscles at another segment (intersegmental reflex) as well as in the original segment (intrasegmental reflex). This divergence to other body regions, common to the flexion reflex, never occurs in the myotatic reflex.

In most mammals, withdrawal of the stimulated limb from noxious stimulus usually leads to contraction of the extensor muscles and relaxation of the flexor muscles of the contralateral limb. Such a pattern is called the crossed extension reflex. This reflex is actually an elaboration of the flexor reflex and possibly functions in two ways. It may support the weight of the body during flexion of the contralateral limb, or secondly, if a continuous noxious stimulus is applied it often flows into rhythmatic alternating stepping, which might be a component of reflex stepping. This mechanism may be viewed as a "wired-in" feature of the spinal cord on which stepping in mammals is based.

The last multisynaptic reflex arc considered is the clasp knife reflex. This reflex is dependent on sensory endings (Golgi tendon organs) which are sensitive to overstretch of extensor muscles. The sensory neurons, as usual, are located in the dorsal root ganglion and pass their axons into the dorsal gray columns where they synapse with internuncial neurons. These neurons facilitate ipsilateral flexor alpha motor neurons and inhibit ipsilateral extensor alpha motor neurons. This inhibition is known as autogenic inhibition, since it both originates from and terminates on the same extensor muscle. This pathway serves as a protective reflex which prevents an extensor muscle from contracting against too great a stretching force.

CEREBELLAR MECHANISMS

The cerebellum, while appearing as a single anatomical feature, has had a long and complex phylogenetic history. It is physiologically organized to carry out at least three functionally complex processes of regulation on a suprasegmental level. It arose in aquatic vertebrates as a center of integration for several sensory modalities involved in the important process of orientation of the body to its environment. With the evolution of terrestrial forms, a new sense modality related to the effects of gravity upon support and posture of the body was added to this orientational data pool. A completely new anatomical feature evolved incorporating this function with already existent vestibular functions. Finally, with the evolution of a new motor system, e.g., pyramidal system, a third function, that of timing voluntary motor actions, was integrated into the earlier functional systems.

With these facts in mind, it is not surprising that the terminology applied to the cerebellum is in a confusing state. Originally, the cerebellum was divided into a medial region, the vermis, and two lateral regions, the cerebellar hemispheres. Each of these areas was further subdivided and named with little regard to origin or functional significance. Unfortunately this inadequate terminology is deeply entrenched in the literature and there is little hope of its being changed. Consideration of both anatomical origin and physiological function of various regions of the brain provides a more useful nomenclature. While no single nomenclature has been proposed to fill this gap, some means of compromise should be reached. Table one lists the most commonly used terms, which are herein organized according to phylogenetic origin and roughly represent the major functional regions.

TABLE 1: DIVISIONS OF THE CEREBELLUM

Flocculo-nodular lobe -	Archicerebellum - Vestibulocerebellum	
	Flocculus Paraflocculus Nodule	Regulation of vestibular reflexes underlying posture.
Corpus cerebelli -	Paleocerebellum - Somatocerebellum	
	Lingula Central lobule Culmen Declive Folium Tuber Pyramid Uvula	Regulation of reflex tonus of postural muscles by inhibition or reinforcement of extensor muscles.
	Neocerebellum - Cerebrocerebellum	
	Ansiform lobule Paramedian lobule	Regulation of corrective reflexes of posture and timing of muscular contractions of voluntary actions.

The archicerebellum, oldest of the cerebellar divisions, evolved in conjunction with the vestibular system which is mainly responsible for gravitiational orientation in higher vertebrates excluding man. Because of this origin and functional similarity, the archicerebellar and vestibular patterns (Diagram 2) will be considered as a single functional system. The afferent pathway is predominantly from the semicircular canals and vestibular portions of the inner ear. These structures relay changes in orientation of the animal's body in all three planes to several secondary centers. The most primitive of these centers are the vestibular nuclei which control postural reflexes of the head and neck. These centers also influence the spinal reflex patterns of posture which were described earlier. While no direct connections have been traced from the vestibular nuclei, it is believed that second order connections are made with the cerebral cortex via the medial geniculate body. Such connections would account for the sensation of vertigo which is perceived at a conscious level.

More complex connections are made with the archicerebellum and its efferent relay nuclei, the fastigial nuclei. Such connections probably represent a higher level of

Diagram 2. Archicerebellar & Vestibular Patterns

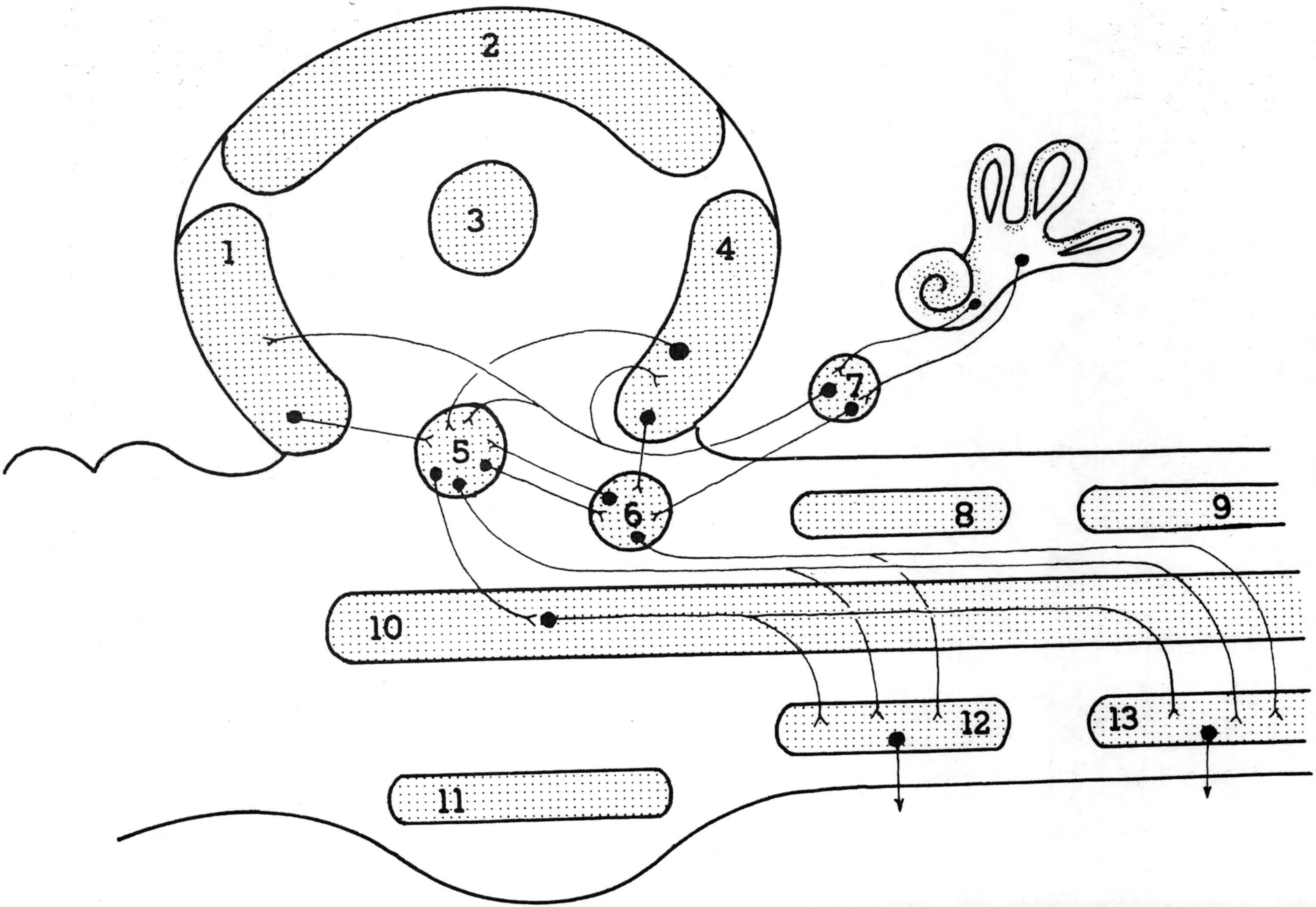

DIAGRAM 2: ARCHICEREBELLAR AND VESTIBULAR PATTERNS

1. Paleocerebellum
2. Neocerebellum
3. Dentate nucleus
4. Archicerebellum
5. Fastigial nucleus
6. Vestibular nuclei
7. Vestibular ganglion
8. Cranial nerve sensory nuclei
9. Dorsal gray column of spinal cord
10. Reticular formation
11. Pontine nuclei
12. Cranial nerve motor nuclei
13. Ventral gray column of spinal cord

regulation of the vestibular reflexes for there are feedback systems from the vestibular nuclei as well as other cerebellar regions. This type of system would allow integration of vestibular sensations with other sensory modalities.

Even at this level of cerebellar organization, a condition is recognized that will hold true for all functional systems of the cerebellum, namely that no direct motor connections are made on the spinal patterns. Instead the cerebellum will function on already existent pathways. The cerebellum will influence these final pathways exclusively by means of the more primitive motor systems of the brain stem (midbrain and medullar centers). Because of this indirect method of action, it is strikingly different from the other major mammalian suprasegmental structure, the cerebral cortex.

With the evolution of terrestrial life and the resultant development of a new mode of locomotion, new problems of orientation of the body arose. It became increasingly important for an animal to know the position of its limbs in order to move them in an organized manner. The problem of body orientation and its importance in locomotion, while significant, is seldom stressed. In a general sense it is possible to view much of the brain's functions as being involved in this problem of orientation.

The sensory endings responsible for this sense of awareness of general body relationships (proprioception) are located throughout the entire body. They include the annulospiral endings of muscles, the pressure endings in connective tissue and skin, receptor organs in joints, the special sensory organs such as eyes, vestibule of the inner ear, and many others. The sensory cell bodies of trunk proprioceptors are located in the dorsal root ganglia and dorsal gray column of the spinal cord. They pass their axons into the lateral funiculus of the cord, where they ascend up the brain stem and are joined by similar connections from the cranial nerves which supply the head and neck. These tracts gather into a great ascending bundle, the posterior cerebellar peduncle or restiform body, and project onto the paleocerebellum where they are joined by projections from the special senses (not represented in Diagram 3).

The integration of vestibular, visual, and general somatic sensory (somesthetic) inputs allow the paleocerebellum to control two phases of locomotion via its efferent connections with the fastigial nuclei and the reticular formation (Diagram 3). Its connections with visual and vestibular inputs supply the raw data for the final adjustment of equilibration and upright posture, a function which to a lesser extent is shared with the archicerebellum. The other function of the paleocerebellum is the regulation of muscle tonus. This regulation is performed by descending fibers acting on both the alpha and gamma motor neurons of the myotatic reflex arc. The general effect of the paleocerebellum is to inhibit the extensor tonus and to facilitate flexor tonus.

The neocerebellum is found only in mammals and closely parallels the evolution of the pyramidal system of the cerebral cortex. The evolution of the pons, which serves as a relay nucleus from the cerebrum to the neocerebellum, is thus closely correlated with its evolution. The chief afferent connections of the neocerebellum originate in the frontal and parietal regions of the cerebral cortex and then pass to the pontine nuclei. After forming synaptic connections with the pontine nuclei, they pass dorsally around the outside of the brain stem forming a massive tract, the middle cerebellar peduncle (classically called the brachium pontis), before projecting onto the neocerebellum.

At present, injuries to the neocerebellum resulting in functional disturbances present the best evidence of its normal functional role. In humans with neocerebellar disturbances and animals with experimental neocerebellar lesions, voluntary movements are slow to start and to stop. In addition, these movements are usually accompanied by lack of direction and steadiness, resulting in overshooting movements by such individuals when reaching for objects. These dysfunctions indicate that the neocerebellum has an important role in timing and guiding of voluntary movements.

This role requires elaborate interconnections, both afferent and efferent, between the cerebral cortex and neocerebellum. Afferent connections have already been described (middle cerebellar peduncle) which might serve as one limb of such a system. The anterior cerebellar peduncle, or brachium conjunctivum, arises from the dentate nucleus, whose afferent fibers are received from the neocerebellar cortex. The middle cerebellar peduncle or tract might serve as the efferent limb of such a system. This peduncle passes anteriorly to the ruber nucleus where part of its fibers synapse and then project upon the lower motor neurons via the reticular formation. The major part of the peduncle, however, continues anteriorly and projects upon the cerebral cortex after forming connections in the thalamic nuclei. This system forms a loop which probably serves as a feedback system between a command center (cerebral cortex) and a stabilizing and timing center (neocerebellum).

SOMATIC SENSORY MECHANISMS

The brain's role as major integrator of the vertebrate body is primarily dependent on two factors: knowledge and coordination of internal and external stimuli. Integration is dependent on the sensory information of body position or proprioception, and the general sensory modalities of pain, temperature, touch, and pressure which are relayed to the brain (Diagram 4). These sensory components are coordinated and compared to past sensory components. This integration results in facilitory or inhibitory motor action on muscles and glands thereby bringing the animal into equilibrium with its environment.

The ascending pathways for pain, temperature, and touch are phylogenetically older than those of tactile discrimination and cerebral proprioception, which arose with the evolution of terrestrial life. The sensory endings of these more ancient pathways vary as to structure but primarily consist of nonmyelinated free nerve endings whose cell bodies are again located in the dorsal root ganglia of spinal nerves. They pass their axons into the dorsal gray column and form synaptic relationships with internuncial neurons. The axons of some internuncial neurons decussate (crossover) via the ventral gray commissure and pass into the lateral funiculus. They then pass forward and are joined by similar pathways from the trigeminal nerve, which forms the main sensory pathway for pain and temperature sensations from the head region. These combined pathways, as well as many others, project onto the ventral nuclei of the thalamus without further primary synaptic relationships. These massive bundles passing through the brain stem are called the lemniscal fibers and are divided into a medial and lateral division, the medial and lateral lemniscus.

On reaching the thalamic nuclei, they again form synapses and project onto the cerebral cortex by axons which pass through the basal nuclei (caudate, putamen, and globus pallidus) as a massive bundle called the internal capsule. On emerging above the basal nuclei, they spray out into the cortex from a fanlike pattern of fibers called the corona radiata. Many of the thalamic fibers, however, do not reach the cerebral cortex but terminate in the basal nuclei where they form connections which are phylogenetically older than those of the cerebral cortex.

With the evolution of terrestrial vertebrate forms, new sensory pathways involved in tactile discrimination and muscle, tendon, and joint sensibility (proprioception) arose. The sensory endings involved in these modalities also have their cell bodies in the dorsal root ganglia but do not pass directly to the thalamic nuclei. The cell bodies of these modalities first pass their axons into the dorsal funiculus where short collateral branches form synaptic relationships with internuncial neurons which in turn synapse on

Diagram 3. Paleocerebellar & Neocerebellar Patterns

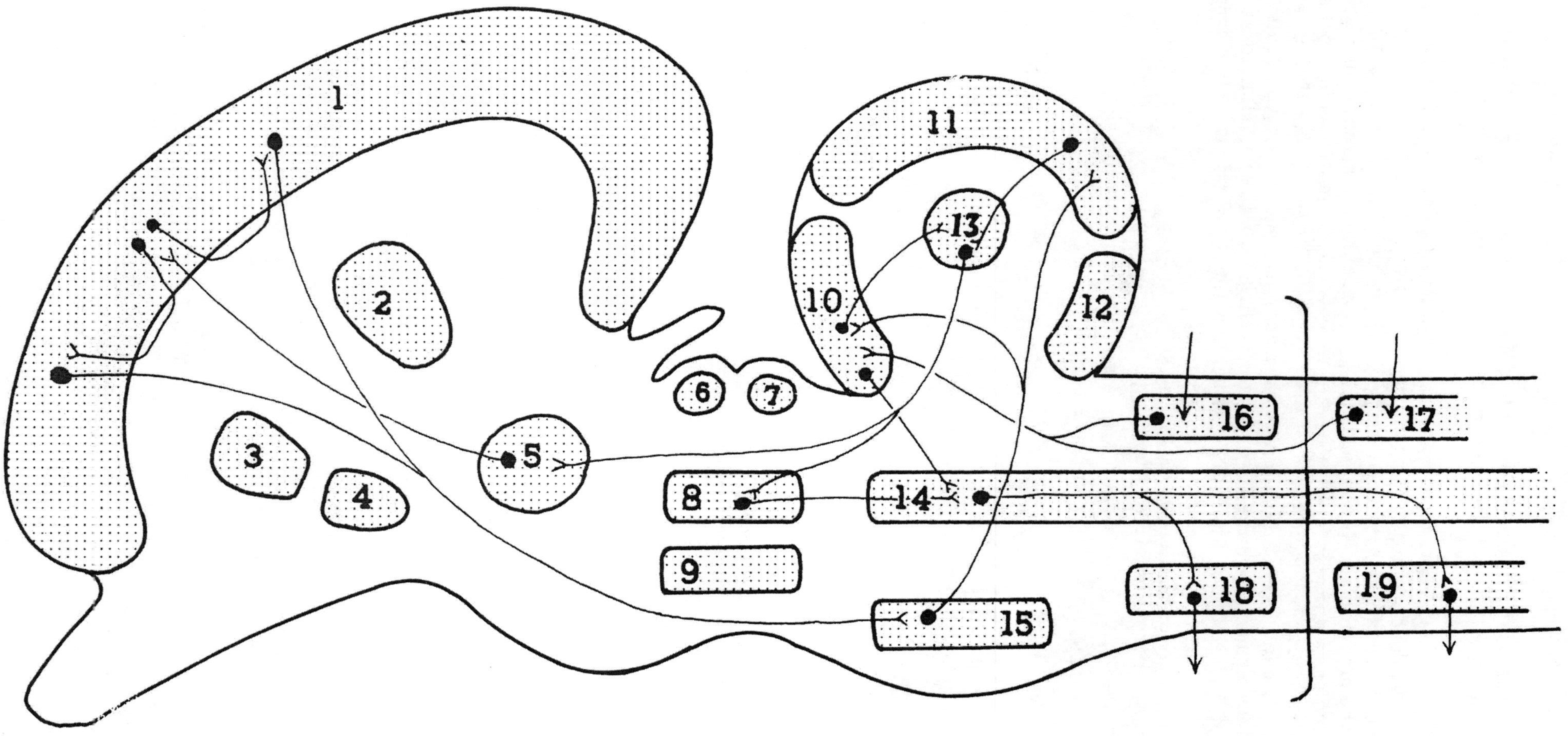

DIAGRAM 3: PALEOCEREBELLAR AND NEOCEREBELLAR PATTERNS

1. Cerebral cortex (neocortex)
2. Caudate nucleus
3. Putamen nucleus
4. Globus pallidus
5. Thalamic nuclei
6. Anterior colliculus
7. Posterior colliculus
8. Ruber nucleus
9. Substantia nigra
10. Paleocerebellum
11. Neocerebellum
12. Archicerebellum
13. Dentate nucleus
14. Reticular formation
15. Pontine nuclei
16. Cranial nerve sensory nuclei
17. Dorsal gray column of spinal cord
18. Cranial nerve motor nuclei
19. Ventral gray column of spinal cord

motor neurons, thereby forming part of the spinal reflexes. The main axons, however, pass up the dorsal funiculus and end in the gracile and cuneate nuclei of the medulla oblongata. Here the second order neurons pass their axons ventrally and cross to the opposite side there they ascend via the lemniscal fibers to the lateral and medial thalamic nuclei. Third order neurons located in these thalamic nuclei pass their axons into the internal capsule where connectors are formed in the basal nuclei and the cerebral cortex (Diagram 4).

In order to integrate sensory modalities and coordinate motor actions, the brain must not only receive information of sensations such as pain and pressure but it must also receive information as to where these sensory modalities are originating.

Each spinal nerve innervates only a segment of the body called a dermatome. Each spinal nerve passes its sensory fibers into the funiculi of the spinal cord; these fibers are not scattered indiscriminately but are added to the external surface of the fasciculi. This results in a laminated tract or fasciculus, representing the dermatomes of the body surface. Such an arrangement of fibers is said to be topographically organized by sensory modality.

Thalamic projection onto the cerebral cortex (primarily the parietal region of the cortex) results in loss of topographical organization according to sensory modality. However, one type of topographical organization is retained in the cortex, according to body region rather than by sensory modality. This may be represented by a lateral view of the body inverted and superimposed on the surface of the cerebral cortex. Various regions of the body represented on the cortex are disproportionately enlarged or reduced according to the quantity of sensory information coming from a given region. Thus the projected head and distal portions of the appendages in man are greatly enlarged whereas other areas of the projected body are reduced. This is not true of all mammals. The regions which supply the greatest amount of sensory information change in various mammals according to their structure and behavior.

RETICULAR MECHANISMS

Anatomically the reticular formation reflects the oldest phylogenetic neuronal pattern on which the vertebrate brain is based. This central core of neurons and fibers (Diagram 4) begins at the medulla oblongata and ascends along the brain stem to the diencephalon, where it joins a series of thalamic nuclei, which probably have evolved from the reticular formation itself. These nuclei are at least functionally associated with this system.

The reticular neuronal pattern consists of a neuropil of short and long-fibered neurons which form multisynaptic connections up and down the reticular formation as well as with surrounding structures. Although this formation has been divided into a large number of nuclei, recent physiological work supports the view that the functional relations of this system are not based on a nucleate pattern. As may be expected, stimulation or ablation of a single nucleus in this formation does not result in any clear-cut dysfunction.

It is believed that the reticular formation arose in vertebrate evolution as a center of sensory and motor integration between spinal and cranial functions associated with increasing importance of encephalization. Such a system would be expected to have an important role in the organization of ascending sensory information and the coordination and modification of descending motor pathways. The results of experimental studies verify such complex functions.

All ascending pathways, both cranial and spinal, pass collateral fibers into the reticular formation as they proceed toward the thalamic nuclei. The reticular formation connects with midline, intralaminar, and thalamic reticular nuclei. These so-called "nonspecific thalamic nuclei" project onto the cerebral cortex and are involved along with the reticular formation in a phenomenon called the arousal response. This sensory system is not topographically organized and seemingly does not carry modality-specific information. Its effect is to facilitate the cortex to ascending sensory information which is modality-specific and which is transmitted by the classical lemniscal system. In addition, this system is involved in a phenomenon called the recruiting response. The number of impulses per unit of time transmitted by the ascending pathways increase and decrease in amplitude as they enter into synaptic relationship to the nonspecific thalamic nuclei. This mechanism may serve an important role in facilitating or inhibiting incoming information.

In addition, the reticular formation greatly affects descending motor pathways and the final spinal patterns on which these motor pathways act. The reticular formation forms connections with the cerebral and cerebellar cortices as well as the phylogenetically older limbic system and subcortical motor systems. These older systems have evolved from and are still functionally part of the reticular formation. The ruber nucleus and the descending motor nuclei of the extrapyramidal systems, which will be covered later, also evolved from the reticular formation and retain connections with these systems. Facilitation and inhibition are mediated at this level as well as at lower centers.

Finally, both extensor facilitory and extensor inhibitory regions are located within the reticular formation, which has a great influence on these final spinal patterns. Cortical, basal nuclear, and cerebellar inhibitory pathways end in the reticular inhibitory regions thereby adding to the inhibition of the gamma motor neurons. Vestibular and reticular facilitory regions, which act to facilitate the gamma motor neurons, complete a mechanism carefully balanced to activate the final common pathways.

SPECIAL SENSORY AND MOTOR MECHANISMS

Olfactory, optic, and auditory sensations traditionally have been classified as special senses in order to emphasize their importance in the origin and evolution of the brain and to call attention to the cranial nerves which are responsible for these sensations, which have a separate phylogenetic origin from the other cranial nerves. The olfactory pathways will not be included in this section, but will be discussed in conjunction with the limbic system with which they form basic functional connections.

The areas of optic and auditory projection (Diagram 5) fall into two divisions. The more ancient tectal regions, located in the roof of the midbrain, and the recently evolved cortical centers. In vertebrates other than mammals, the highest centers of sensory and motor coordination are located in the tectum of the midbrain and the basal regions of the cerebral hemispheres or telencephalon. With the evolution of mammals, optic and auditory sensations begin to play a more important role and their centers shift into the telencephalon along with the other sensory centers. The older centers and pathways are not lost but retain only a remnant of their former functional importance.

Auditory impulses are relayed from the organ of corti in the inner ear to the dorsal and ventral cochlear nuclei via the spiral ganglion of the cochlear nerve. From these nuclei, connections are formed which are the oldest auditory connections in the vertebrate brain. Fibers project to the spinal cord and cranial motor nuclei via the

Diagram 4 Somatic Sensory & Reticular Patterns

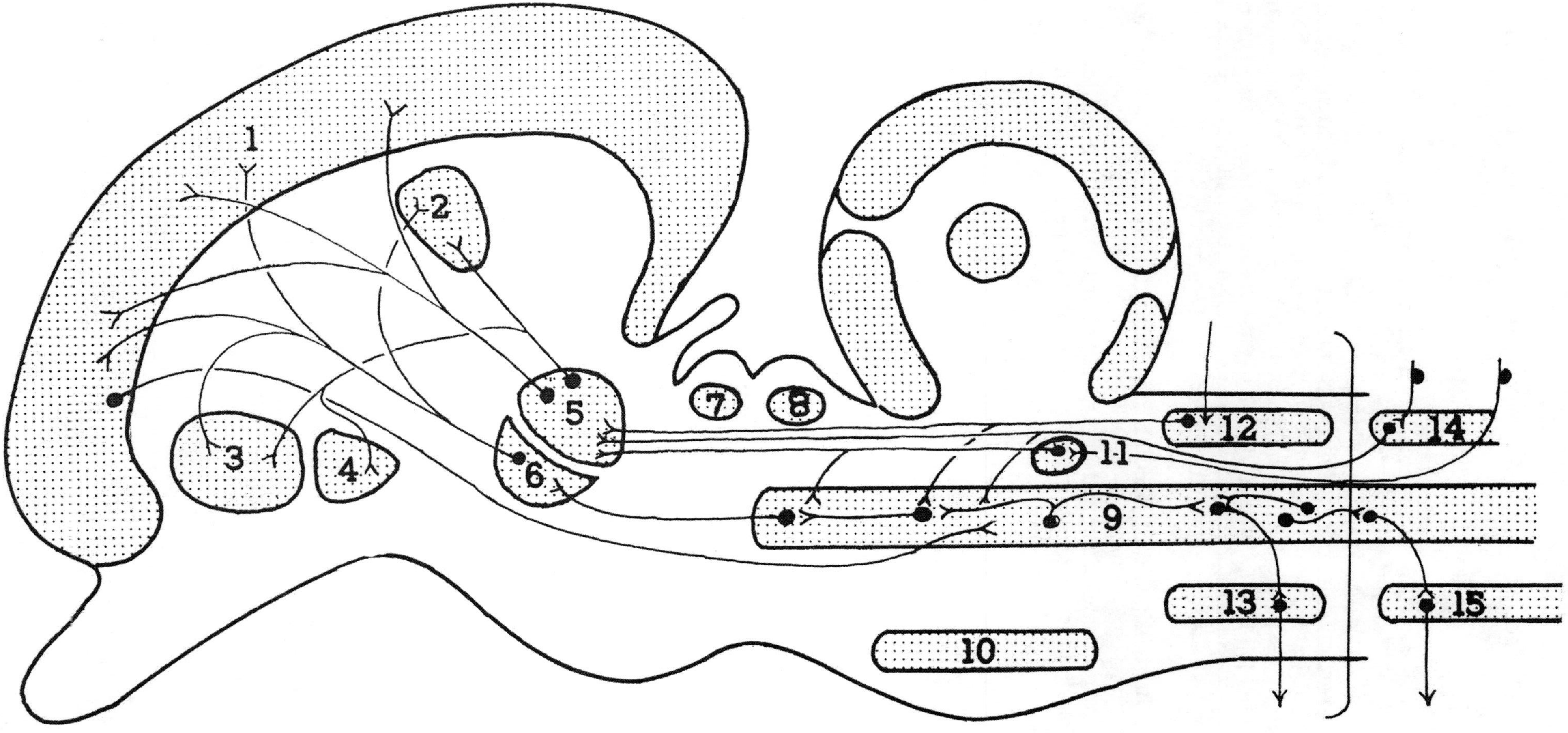

DIAGRAM 4: SOMATIC SENSORY AND RETICULAR PATTERNS

1. Cerebral cortex (neocortex)
2. Caudate nucleus
3. Putamen nucleus
4. Globus pallidus
5. Medial and lateral thalamic nuclei
6. Thalamic reticular nuclei
7. Anterior colliculus
8. Posterior colliculus
9. Reticular formation
10. Pontine nuclei
11. Cuneate and gracile nuclei
12. Cranial nerve sensory nuclei
13. Cranial nerve motor nuclei
14. Dorsal gray column of spinal cord
15. Ventral gray column of spinal cord

Diagram 5 Special Sensory & Motor Patterns

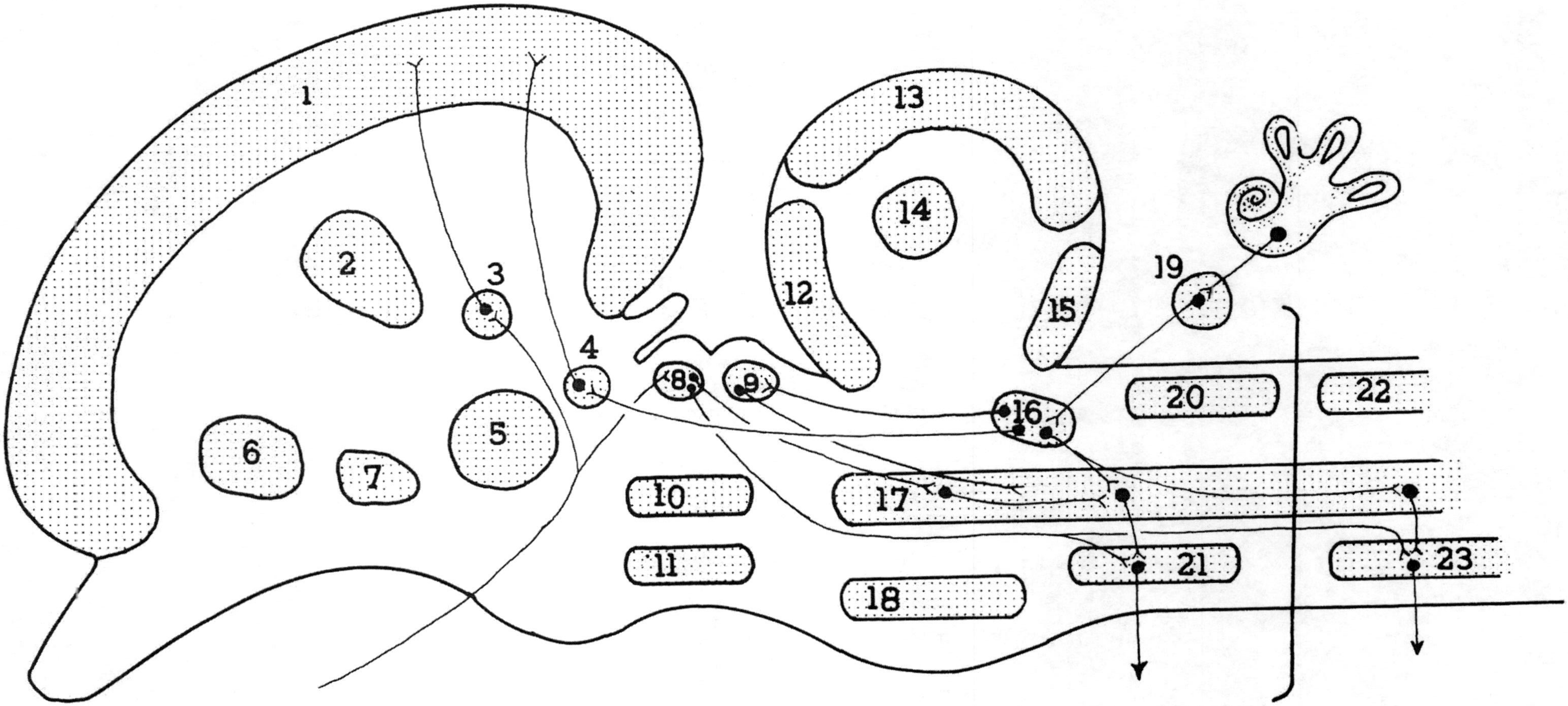

DIAGRAM 5: SPECIAL SENSORY AND MOTOR PATTERNS

1. Cerebral cortex (neocortex)
2. Caudate nucleus
3. Lateral geniculate body
4. Medial geniculate body
5. Thalamic nuclei
6. Putamen nucleus
7. Globus pallidus
8. Anterior colliculus
9. Posterior colliculus
10. Ruber nucleus
11. Substantia nigra
12. Paleocerebellum
13. Neocerebellum
14. Dentate nucleus
15. Archicerebellum
16. Dorsal and ventral cochlear nuclei
17. Reticular formation
18. Pontine nuclei
19. Spiral ganglion
20. Cranial nerve sensory nuclei
21. Cranial nerve motor nuclei
22. Dorsal gray column of spinal cord
23. Ventral gray column of spinal cord

the reticular formation. Connections are also formed with the posterior colliculi of the corpora quadrigeminia or tectum as it is called in lower vertebrates. These connections were first established in amphibians where they provide the highest center of sensory coordination. In mammals this region is much reduced and functions only in auditory reflexes such as the turning of the pinna and the "startle response" to sudden loud noises.

The pathways involved in cortical auditory sensations by-pass these more ancient centers. These sensations are projected to the temporal region of the cerebral cortex by the lateral lemniscus, which begins in the cochlear nuclei and projects onto the cortex via the medial geniculate body. The medial geniculate body is one of several posterior nuclei of the dorsal thalamus which become important with the vast increase in functional importance of the telencephalon in mammals. The pathways as traced here seem quite simple, but it must be kept in mind that only major connections have been discussed and that at almost every point, decussation of fibers between the two sides occur. These cross connections are of importance in closely integrating the separate functions of each side, and are even of greater importance in the function of auditory localization, the spatial location of a sound source, which requires the integration of both auditory receptors.

The evolutionary history of optic pathways is quite similar to that of the auditory pathways. The more ancient connections are made with the anterior colliculi of the tectum. From these centers, connections are formed with the cranial and spinal motor nuclei via the reticular formation. These connections give rise to optic reflexes such as turning of the head and protection of the eyes by blinking in response to a sudden bright light. Direct connections from the anterior colliculi to the cranial motor nuclei also exist. These connections are responsible for such reflexes as the photopupil reflex which is involved in regulation of pupil size in response to intensity of light.

Cortical connections are established via the lateral geniculate body. This nucleus is also a member of the posterior nuclear group of the dorsal thalamus and projects to the occipital portion of the cerebral cortex, where visual impulses are integrated with other sensory modalities. As in the case of auditory pathways, considerable decussation between the two sides of the brain is necessary for the functional integration of the visual system.

LIMBIC MECHANISMS

The telencephalon, of which the limbic system is a part, has a complex phylogenetic history about which very little is known. Its structure has undergone great modification and with these changes are mirrored ever increasing functional complexity. This evolution is perhaps best understood as a series of changes from the neuronal fields involved in olfactory and visceral coordination to the neuronal fields partially or totally invaded by somatic coordination functions. In lower aquatic forms, the telencephalon is wholely concerned with olfactory sensations and their coordination with lower visceral centers (habenular, hypothalamic, and midbrain regions). This coordination results in the modulation of feeding and other homeostatic instincts (mating, maternal, and internal homeostatic behavior). With the evolution of terrestrial vertebrates, the telencephalon becomes involved in coordination of general somatic sensations and begins a complex series of changes which led to its present condition.

At present, our neurophysiological knowledge of lower vertebrates is so lacking that little is known of the causal factors which influenced the rapid evolution of the telencephalon. Once vertebrate life was freed from the necessity of returning to an aquatic environment, a vast increase in the range of sensory data that might be experienced resulted. This increase in habitat accompanied by the increase in impinging sensory data might have served as one factor in the change of the structure of the telencephalon. Up to this time, the tectum had served as the primary center of somatic coordination. It is possible that, with this increase in territorial variation in addition to certain mechanical limitations of the already specialized tectal regions, the way was open for the major evolution of new unspecialized coordination regions which possessed a more flexible neuronal pattern.

The evolution of the telencephalon is characterized by the appearance of fields or loci (Table 2) involved in the differentiation of various functional systems. These systems may be divided into olfactory-visceral and somatic-locomotive coordination, but it should be realized that their evolution followed separate paths. The cerebrum has not undergone specialization along a single line but has been involved in many evolutionary lines. Each line is the result of interactions of many environmental factors with structurally complex organisms, some of which have successfully maintained their relationship to an everchanging habitat.

This present discussion is made difficult by the mass of material and complexity of connections confronting it than those of previous sections. To describe adequately the structures and functions of the limbic system would require a text many times greater than the size of the present one. With these limitations in mind, the following discussion should be understood to be an attempt to establish some order in the vast amount of conflicting material, and not to describe a system whose evolution and functions are clearly established.

The olfactory-visceral coordination system, or limbic system, represents the oldest functional system of the telencephalon. Its centers, or loci, like the other cerebral centers, have undergone a complex series of changes. This evolution will be presented as having occurred in three stages.

The first stage is reflected by the cerebral structure of aquatic vertebrates, where the floor (basal) and roof (pallial) centers of the telencephalon receive olfactory sensations from the nasal regions via the olfactory bulbs and tracts (lateral and medial olfactory tracts) as shown in Diagram 6. In these forms, the cerebral centers serve as relay regions for the distribution of olfactory sensations to the diencephalon (interbrain) and the mesencephalon (midbrain). In these centers, coordination with somatic sensations occurs and the entire behavior of the animal is established at a midbrain or tectal level.

In the evolution of the second stage of the limbic system, the primary centers of olfaction are limited to the anterior basal, and the dorsomedial and ventrolateral pallial regions of the telencephalon (corticomedial division of the amygdaloid complex, olfactory tubercle, pyriform cortex, and other allocortical structures). These primary centers form connections with secondary centers, both basal and pallial (the basolateral division of the amygdaloid complex, septal nuclei, frontotemporal and subcallosal cortices). These centers also form connections with each other (anterior commissure, stria medullaris, and stria terminalis) and with diencephalic and tectal regions (hypothalamus, habenula, thalamus, and the floor of the midbrain) as shown in Diagram 6. Visceral and emotional functions are mediated thru these secondary limbic centers, which represent a phylogenetic shift or movement of behavioral centers into the telencephalon. Destruction of these limbic centers results in emotional placidity (amygdaloid complex), hyper-emotionality, such as extreme fear and aggression (septal nuclei), and lack of regulation of motivational behavior (frontotemporal cortex), thus clearly indicating their involvement as behavioral components.

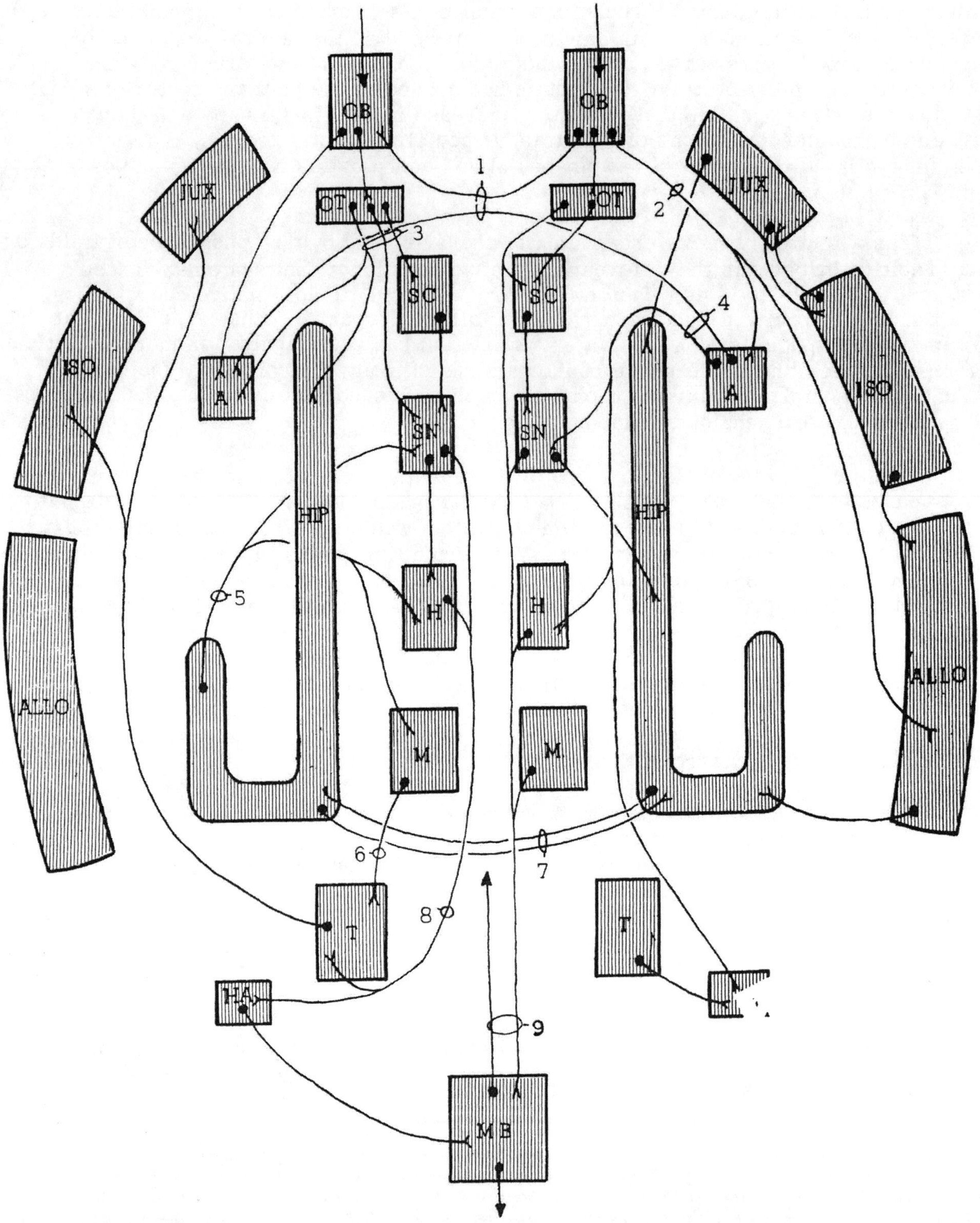

Diagram 6 - Limbic Patterns (dorsal view)

DIAGRAM 6: LIMBIC PATTERNS

A - Amydaloid nuclei

ALLO - Allocortex

H - Hypothalamus

HA - Habenula

HIP - Hippocampus

ISO - Isocortex

JUX - Juxtallocortex

M - Mammillary body

MB - Midbrain

OB - Olfactory bulb

OT - Olfactory tubercule

SC - Subcallosal cortex

SN - Septal nuclei

T - Thalamic nuclei

1. Anterior commissure
2. Lateral olfactory tract
3. Medial olfactory tract
4. Stria terminalis
5. Fornix
6. Mammillothalamic tract
7. Hippocampal commissure
8. Stria medullaris
9. Medial forebrain bundle

TABLE 2: CEREBRAL STRUCTURES

PALLIAL FIELDS

Allocortex - Paleopallium

- entorhinal cortex
- olfactory tubercle
- prepyriform cortex
- pyriform cortex

Archipallium

- hippocampus (Ammon's formation)

Juxtallocortex - Mesopallium

- cingulate cortex
- frontotemporal cortex
- retrosplenial cortex
- subcallosal cortex

Isocortex - Neopallium

- all six-layered cortex
- claustrum

BASAL FIELDS

Limbic subcortical nuclei

- amygdaloid complex (Archistriatum)
  - corticomedial group
  - basolateral group
- area of diagonal band
- olfactory bulbs
- septal nuclei

Nonlimbic subcortical nuclei (Corpus striatum or basal nuclei)

Paleostriatum

- globus pallidus

Neostriatum

- caudate nucleus
- putamen nucleus

The third stage of limbic evolution involves the coordination at a telencephalic level of somatic sensations with the complex olfactory-visceral system. This coordination occurs in tertiary limbic centers (hippocampus, cingulate, entorhinal, and retrosplenial cortices), which have closely evolved with the earlier pallial limbic structures (allocortex and limbic subcortical nuclei) and somatic telencephalic centers (isocortex and nonlimbic subcortical nuclei). Those tertiary centers (juxtallocortex and partially allocortex), intermediate in function between the olfactory-visceral and somatic-locomotive systems, have functions relating to both systems and form connections whereby both systems can influence and coordinate their individual actions as well as their collective actions. The short nonmyelinated tracts which interconnect these centers form elaborate interconnections with each other as well as with lower visceral centers by means of many well established pathways, primarily the fornix, thalamo-cingulate, thalamoisocortical, and medial forebrain bundle projection systems.

SOMATIC-LOCOMOTIVE MECHANISMS

The motor patterns of striated muscles resulting in locomotion vary from one vertebrate to another, a variation which is reflected in the neural areas serving as the highest motor centers. As locomotion becomes more complex, the centers which are involved also become more complex.

In lower aquatic vertebrates, the major correlation center for locomotive integration is the ventrolateral basal regions of the telencephalon (paleostriatum, or globus pallidus). In these animals, swimming movements occur by lateral undulations of the trunk which are characterized by constant rhythmical occurrence of apparently automatic nature.

The mode of locomotion changed with the evolution of terrestrial vertebrates. The limbs, which now supported the body, also provided the means of locomotion. The sway of the trunk, common in amphibians, now became relatively inefficient. Early in the history of terrestrial vertebrate evolution, inhibition of the rhythmical contraction of trunk musculature was brought about by the evolution of the neostriatum (caudate and putamen nuclei). This, however, was not the sole function of the neostriatum. With the forward shift of somatic sensory centers into the telencephalon, additional connections were formed with the neostriatum. In reptiles and birds, these regions are the final sensory and motor centers concerned with the formation of fundamental movement and stereotyped behavior patterns. The sensory projection pathways to these centers are similar in all terrestrial vertebrates (Diagram 7). Connections are formed from the thalamus to the striatal centers, which in turn project back on the thalamus, thus forming a feedback mechanism. In addition, pathways are formed from the striatal centers to the ventral and subthalamic nuclei (ruber nucleus and substantia nigra) and thence to the motor regions of the spinal cord via the reticular formation. Finally connections are also formed with these basal centers and the cerebellum via the reticular and thalamic systems which are clearly demonstrated by means of experimental studies in involvements of motor coordination.

The evolution of the mammalian motor system is characterized by a shift of the dominating motor regions from the striatal centers to cortical centers. This evolution of motor centers has occurred in two stages. The earliest stage is characterized by motor connections from the cerebral cortex (isocortex, or neocortex) to the basal, ventral thalamic, subthalamic, and reticular systems. These changes bring these basal centers under cortical influence and thereby serve to integrate the earlier centers with the more recent certical motor functions. The new functional system created by this fusion, which is found even in the most primitive mammals, is referred to as the extra-pyramidal motor system (Diagram 7).

Diagram 7 Pyramidal & Extrapyramidal Patterns

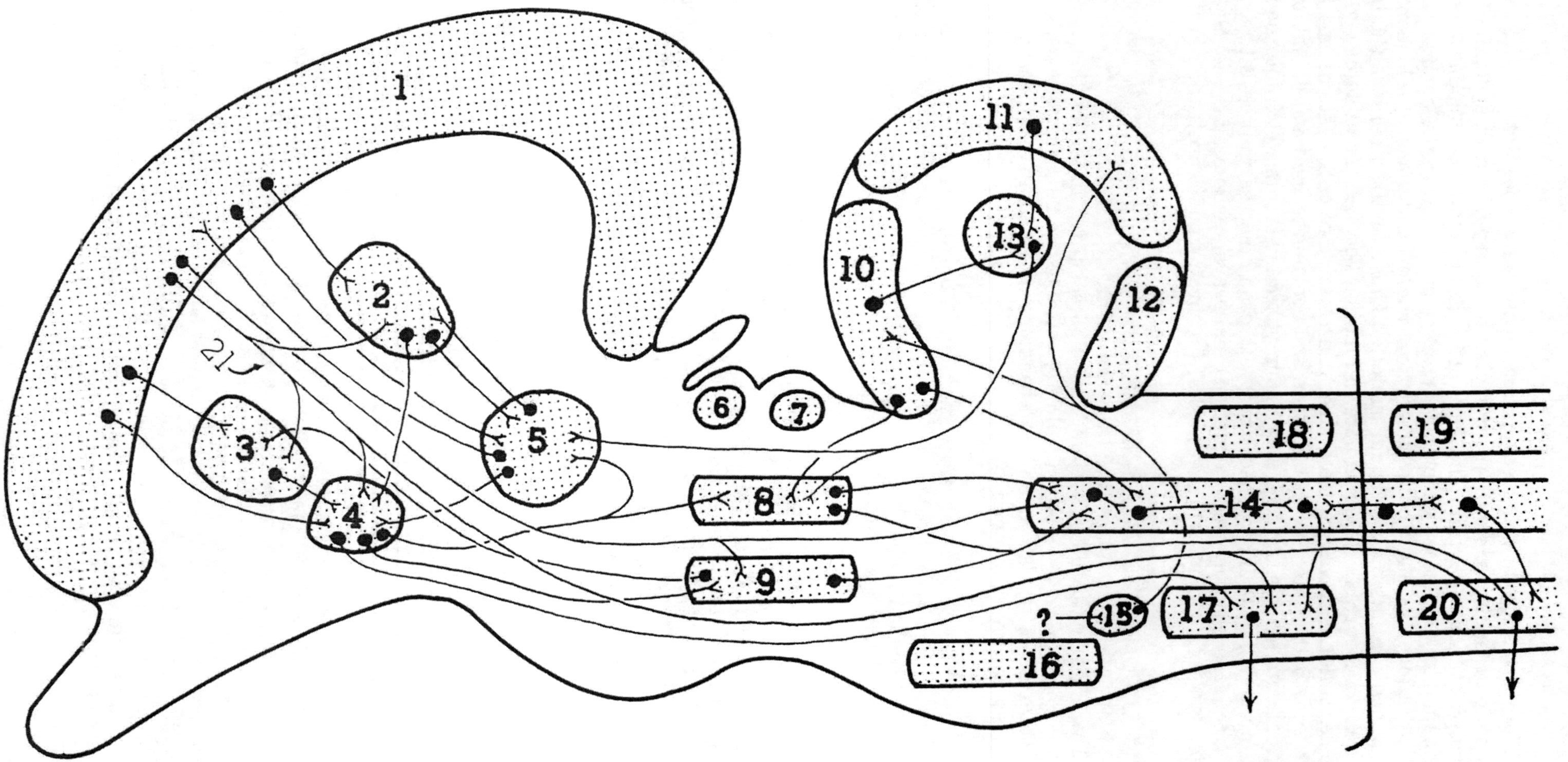

DIAGRAM 7: EXTRAPYRAMIDAL AND PYRAMIDAL MOTOR PATTERNS

1. Cerebral cortex (neocortex)
2. Caudate nucleus
3. Putamen nucleus
4. Globus pallidus
5. Thalamic nuclei
6. Anterior colliculus
7. Posterior colliculus
8. Ruber nucleus
9. Substantia nigra
10. Paleocerebellum
11. Neocerebellum
12. Archicerebellum
13. Dentate nucleus
14. Reticular formation
15. Inferior olive nucleus
16. Pontine nuclei
17. Cranial nerve motor nuclei
18. Cranial nerve sensory nuclei
19. Dorsal gray column of spinal cord
20. Ventral gray column of spinal cord
21. Pyramidal (corticospinal) tract

The second stage in the evolution of mammalian motor centers is characterized by the development of a heavily myelinated cortical tract which by-passes the basal centers and passes directly to the cranial and spinal motor centers. This system is referred to as the pyramidal, or corticospinal tract, and in primates is concerned with finely graded motor movements of the face, hands, and feet.

The evolution of these cortical systems (extrapyramidal and pyramidal systems) has greatly changed the structure of the telencephalon. Connections from these systems to lower brain stem structures divides the neostriatum into a medial caudate nucleus and a lateral putamen nucleus. This places a portion of the neostriatum (putamen) with the paleostriatum (globus pallidus), which together are referred to as the lentiform nucleus. The passage of cortical fibers through these basal centers help create the internal capsule, a prominent feature in all mammalian brains.

These systems, while evolving at different times, are marvelously integrated to coordinate and affect the motor control carried out by all mammals. Experimental studies show that these systems do not function independently of each other but are integrated in a complex manner. If the cortex is ablated, stimulation of striatal centers does not result in movement and yet stimulation of these centers in an intact animal causes the inhibition of cortically induced movements. These striatal areas must modulate locomotive activity since destruction of them leads to dysfunction of locomotive activity. It is suggested that the striatal centers have separate functions which act synergistically to modulate higher cortical motor activity. The exact manner of this modulation remains one of the many mysteries of the evolution and function of the nervous system.

III. BIBLIOGRAPHY

Briggs, E. A. 1939 Anatomy of the Sheep's Brain. Angus and Robertson Limited, Sydney.

Crosby, E. C., T. Humphrey and E. W. Lauer 1962 Correlative Anatomy of the Nervous System. Macmillan, New York

Herrick, C. Judson 1948 The Brain of the Tiger Salamander. University of Chicago Press.

Ariens Kappers, C. U., G. C. Huber and E. C. Crosby 1936 The Comparative Anatomy of the Nervous System of Vertebrates Including Man. Two vols, Macmillan, New York. (Reprinted in 1960 by Hafner, New York in 3 vols.).

Krieg, W. J. S. 1953 Functional Neuronantomy. Blakiston Company, New York.

Papez, J. W. 1929 Comparative Neurology. Crowell, New York (Reprinted in 1960 by Hafner, New York).

Pribram, K. H., and Lawrence Kruger 1954 Functions of the "olfactory brain". Annals of N. Y. Acad. Sci. 58:109-138.

Rasmussen, A. T. 1952 The Principal Nervous Pathways. Macmillan, New York.

Romer, A.S. 1956 The Vertebrate Body, Saunders, Philadelphia.

Ruch, T. C., and J. F. Fulton 1960 Medical Physiology and Biophysics. Saunders, Philadelphia.

Schepers, G. W. H. 1948 Evolution of the Forebrain. Maskew Miller Limited, Cape Town.

Smith, H. M. 1960 Evolution of Chordate Structure. Holt, Rinehart and Winston, Inc., New York.

Wilkie, J. 1937 The Dissection and Study of the Sheep's Brain. Oxford University Press, London.

Zeman, W., and J. R. M. Innes 1963 Craigie's Neuroanatomy of the Rat. Academic Press, New York.

IV. ATLAS

Figure 1

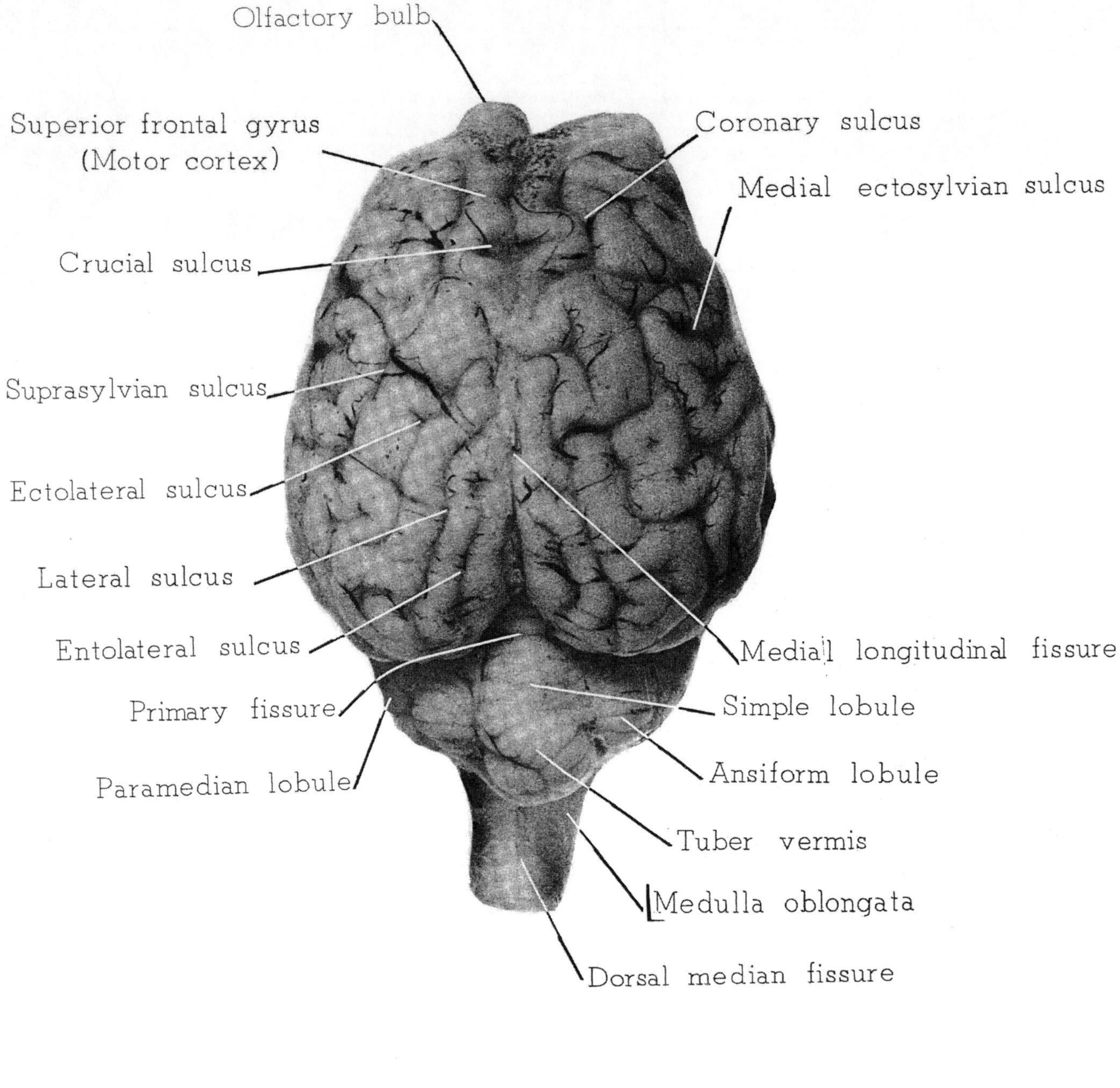
Olfactory bulb
Superior frontal gyrus
(Motor cortex)
Coronary sulcus
Medial ectosylvian sulcus
Crucial sulcus
Suprasylvian sulcus
Ectolateral sulcus
Lateral sulcus
Entolateral sulcus
Medial longitudinal fissure
Primary fissure
Simple lobule
Paramedian lobule
Ansiform lobule
Tuber vermis
Medulla oblongata
Dorsal median fissure

Figure 2

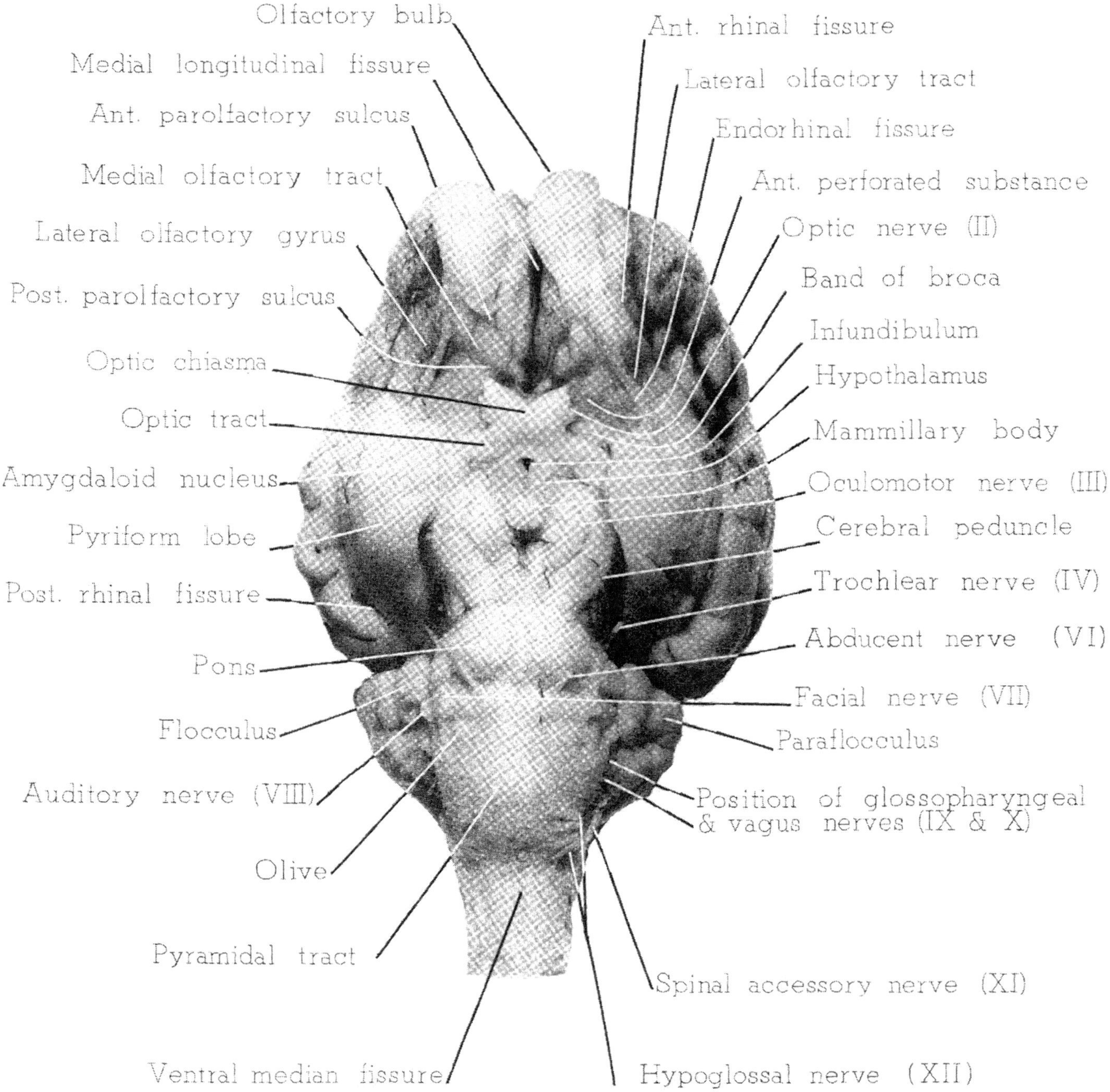

Olfactory bulb
Medial longitudinal fissure
Ant. parolfactory sulcus
Medial olfactory tract
Lateral olfactory gyrus
Post. parolfactory sulcus
Optic chiasma
Optic tract
Amygdaloid nucleus
Pyriform lobe
Post. rhinal fissure
Pons
Flocculus
Auditory nerve (VIII)
Olive
Pyramidal tract
Ventral median fissure
Ant. rhinal fissure
Lateral olfactory tract
Endorhinal fissure
Ant. perforated substance
Optic nerve (II)
Band of broca
Infundibulum
Hypothalamus
Mammillary body
Oculomotor nerve (III)
Cerebral peduncle
Trochlear nerve (IV)
Abducent nerve (VI)
Facial nerve (VII)
Paraflocculus
Position of glossopharyngeal
& vagus nerves (IX & X)
Spinal accessory nerve (XI)
Hypoglossal nerve (XII)

Figure 3

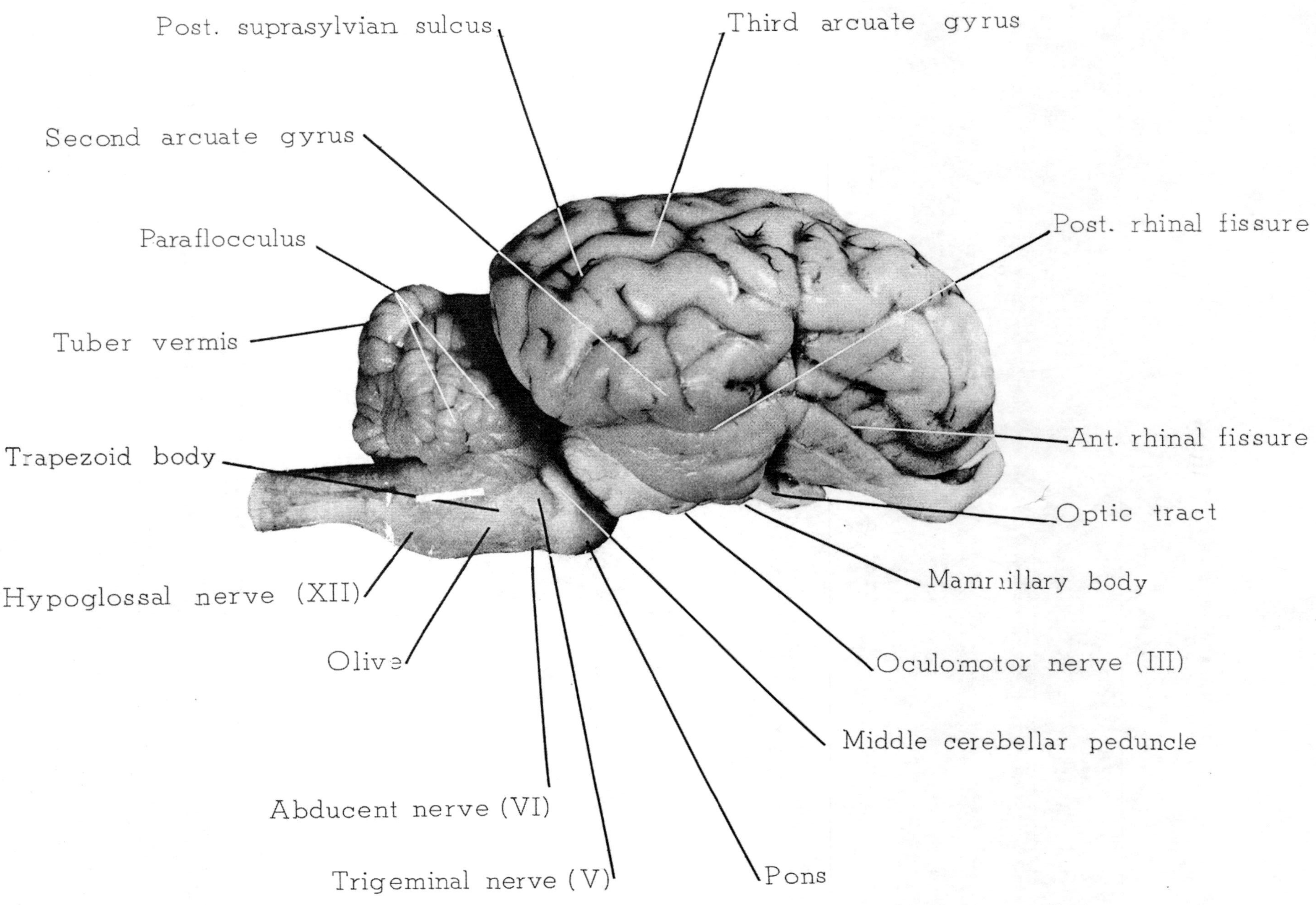
Post. suprasylvian sulcus
Third arcuate gyrus
Second arcuate gyrus
Paraflocculus
Post. rhinal fissure
Tuber vermis
Ant. rhinal fissure
Trapezoid body
Optic tract
Mamrıillary body
Hypoglossal nerve (XII)
Olive
Oculomotor nerve (III)
Middle cerebellar peduncle
Abducent nerve (VI)
Trigeminal nerve (V)
Pons

Figure 4

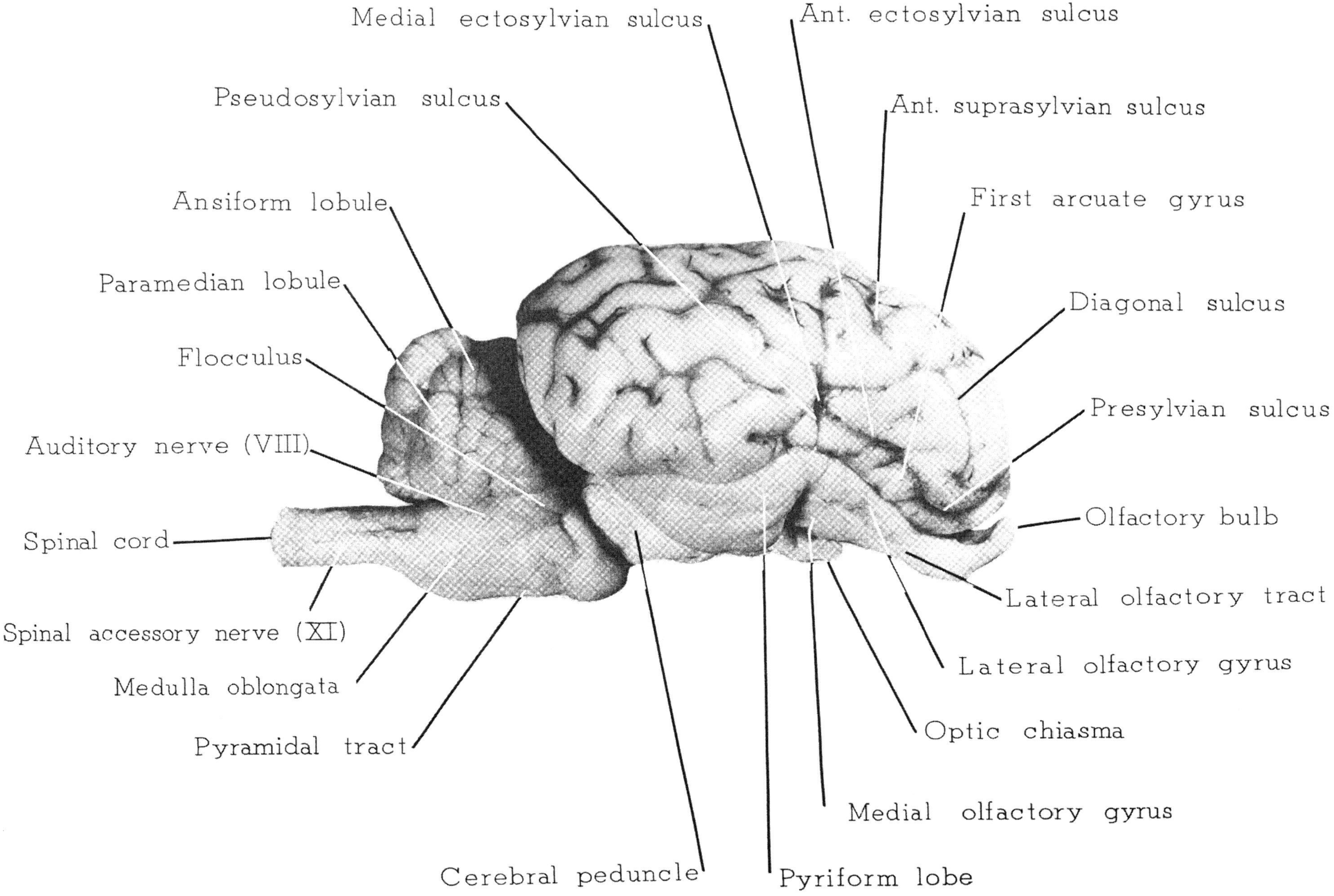

Medial ectosylvian sulcus
Ant. ectosylvian sulcus
Pseudosylvian sulcus
Ant. suprasylvian sulcus
Ansiform lobule
First arcuate gyrus
Paramedian lobule
Diagonal sulcus
Flocculus
Presylvian sulcus
Auditory nerve (VIII)
Olfactory bulb
Spinal cord
Lateral olfactory tract
Spinal accessory nerve (XI)
Medulla oblongata
Lateral olfactory gyrus
Pyramidal tract
Optic chiasma
Medial olfactory gyrus
Cerebral peduncle
Pyriform lobe

Figure 5

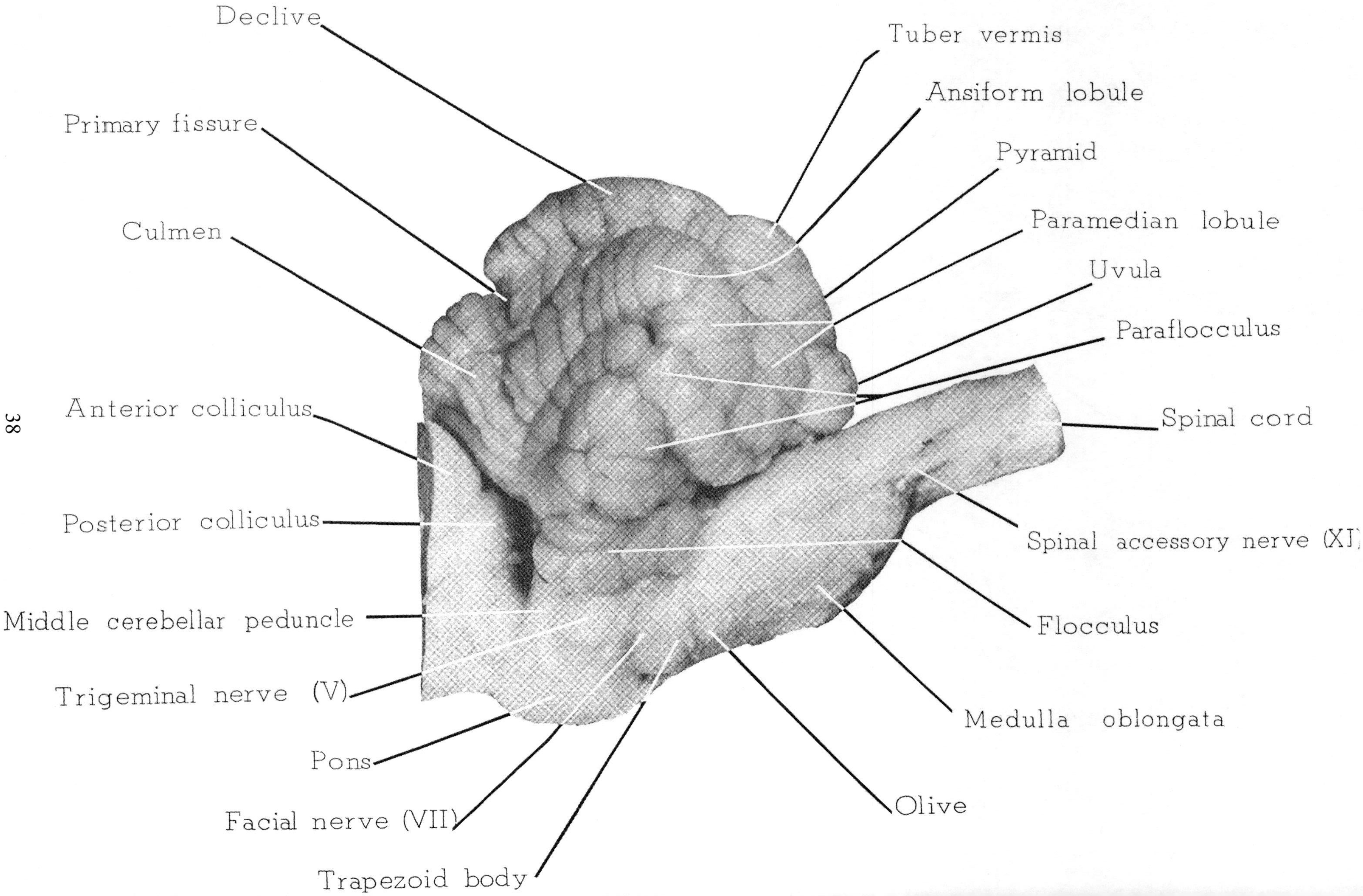
Declive
Primary fissure
Culmen
Anterior colliculus
Posterior colliculus
Middle cerebellar peduncle
Trigeminal nerve (V)
Pons
Facial nerve (VII)
Trapezoid body
Tuber vermis
Ansiform lobule
Pyramid
Paramedian lobule
Uvula
Paraflocculus
Spinal cord
Spinal accessory nerve (XI)
Flocculus
Medulla oblongata
Olive

Figure 6

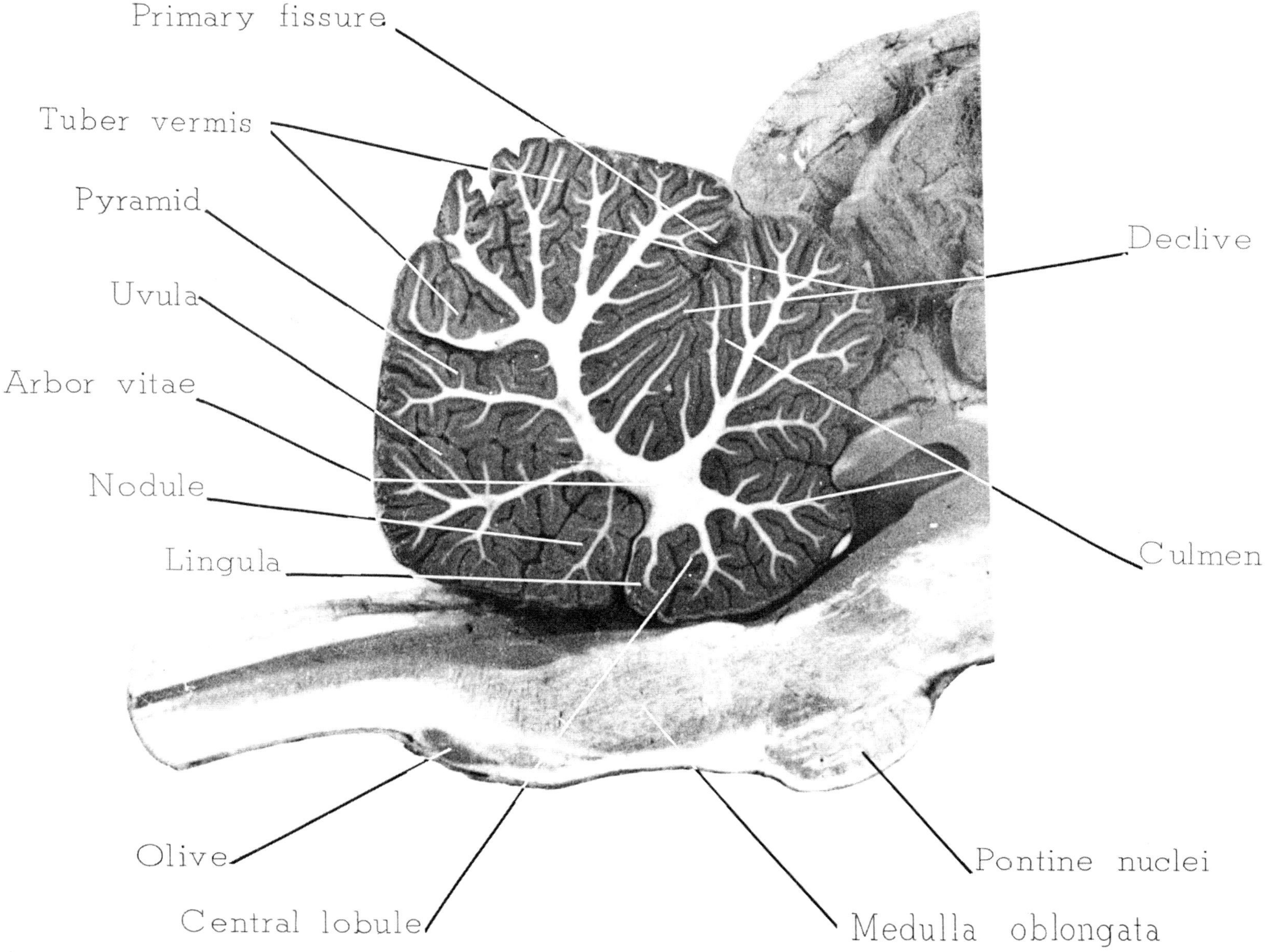
Primary fissure
Tuber vermis
Pyramid
Uvula
Arbor vitae
Nodule
Lingula
Declive
Culmen
Olive
Central lobule
Pontine nuclei
Medulla oblongata

Figure 7

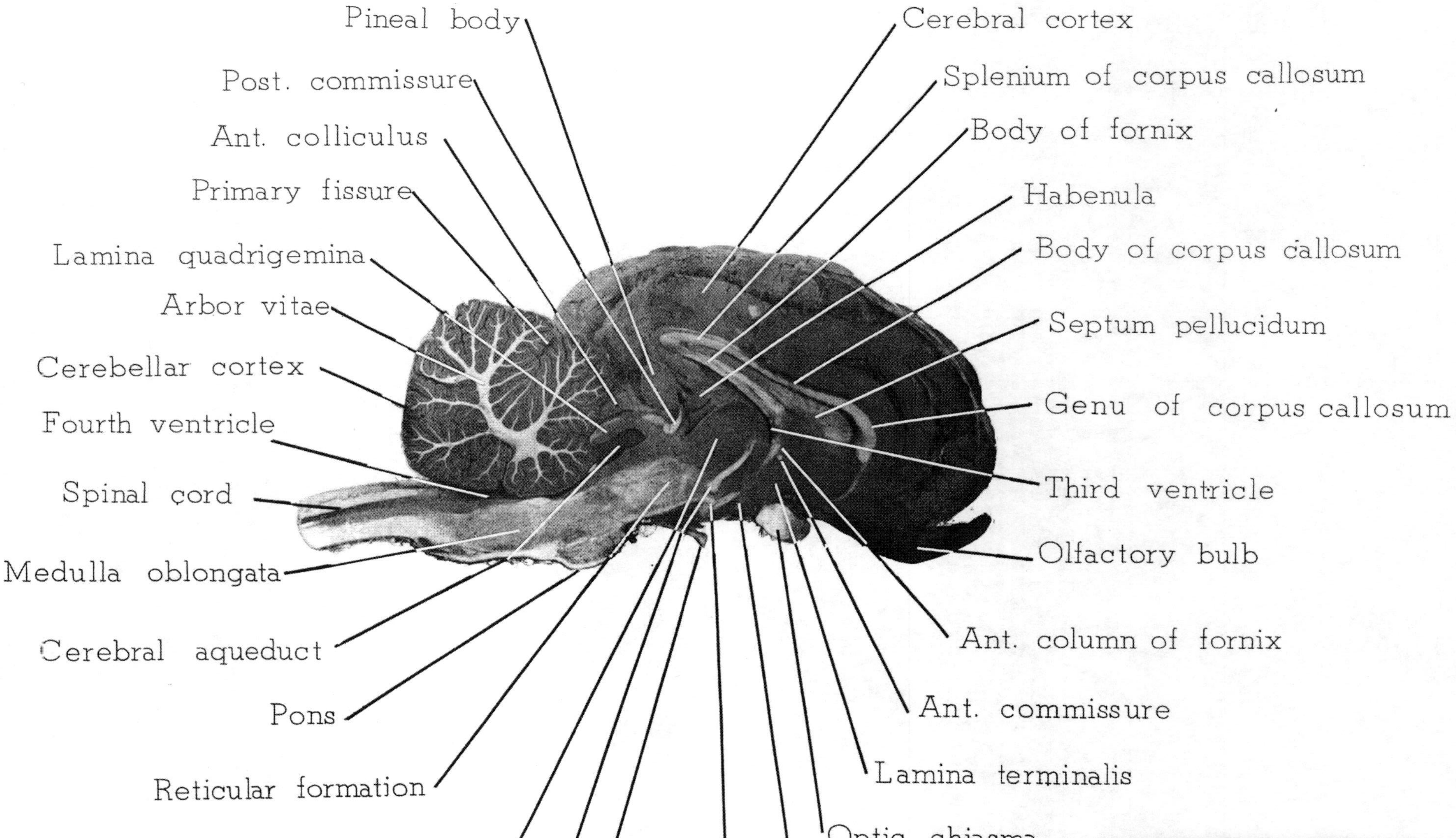

Pineal body
Post. commissure
Ant. colliculus
Primary fissure
Lamina quadrigemina
Arbor vitae
Cerebellar cortex
Fourth ventricle
Spinal cord
Medulla oblongata
Cerebral aqueduct
Pons
Reticular formation
Intermediate mass of thalamus
Mammillothalamic tract
Oculomotor nerve (III)
Cerebral cortex
Splenium of corpus callosum
Body of fornix
Habenula
Body of corpus callosum
Septum pellucidum
Genu of corpus callosum
Third ventricle
Olfactory bulb
Ant. column of fornix
Ant. commissure
Lamina terminalis
Optic chiasma
Hypothalamus
Mammillary body

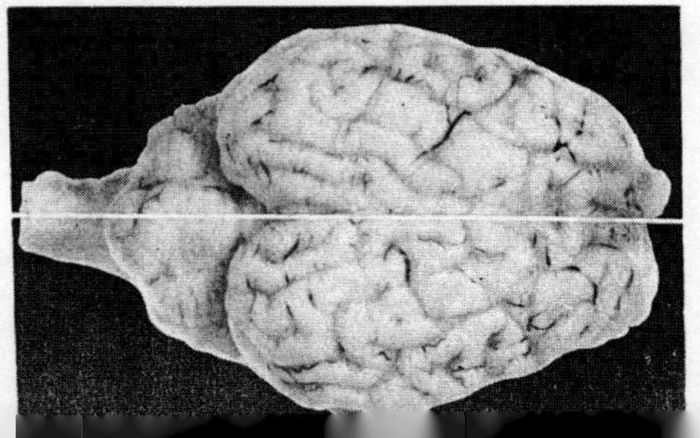

Figure 8

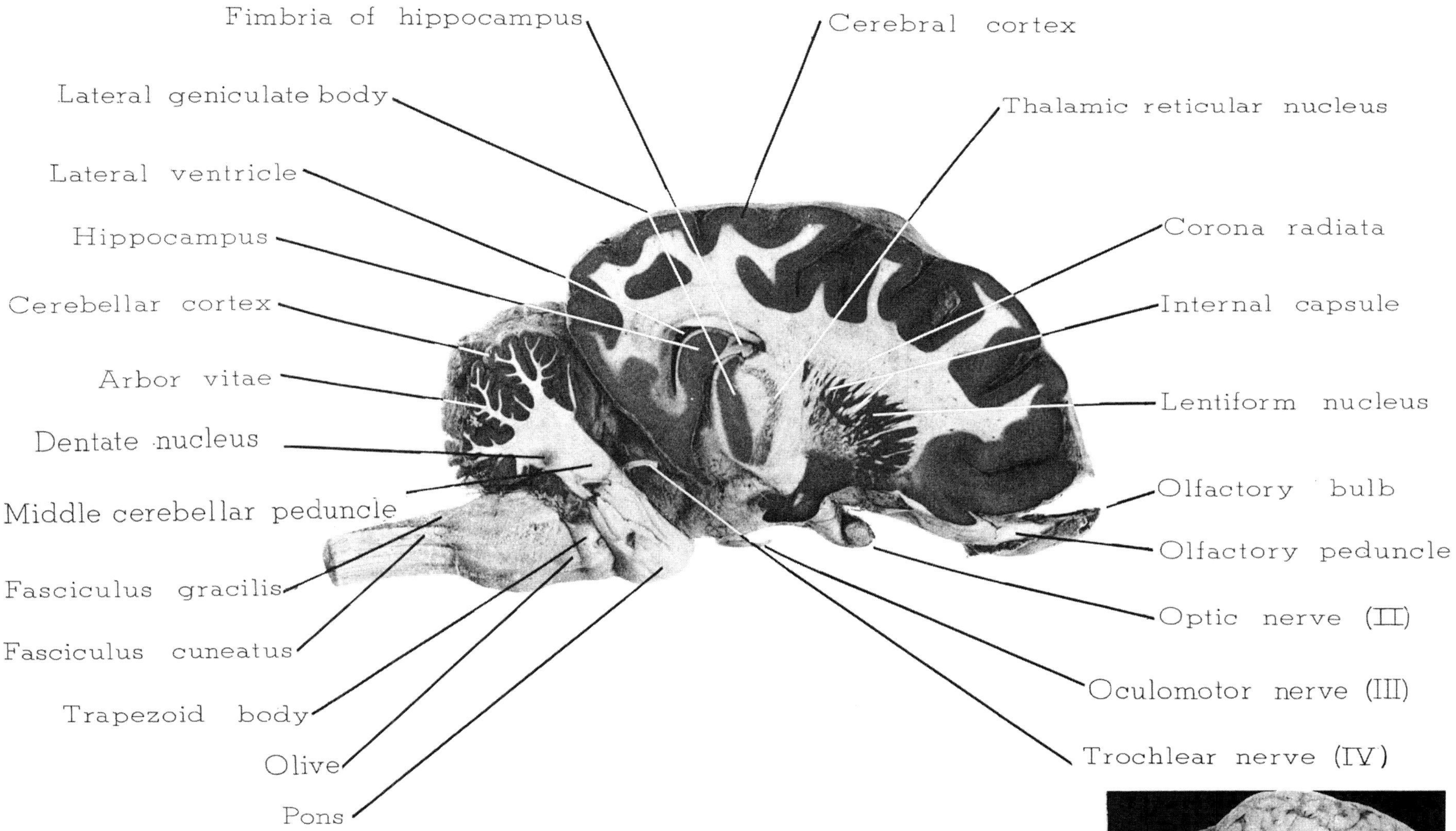

Fimbria of hippocampus
Cerebral cortex
Lateral geniculate body
Thalamic reticular nucleus
Lateral ventricle
Hippocampus
Corona radiata
Cerebellar cortex
Internal capsule
Arbor vitae
Lentiform nucleus
Dentate nucleus
Middle cerebellar peduncle
Olfactory bulb
Olfactory peduncle
Fasciculus gracilis
Fasciculus cuneatus
Optic nerve (II)
Trapezoid body
Oculomotor nerve (III)
Olive
Trochlear nerve (IV)
Pons

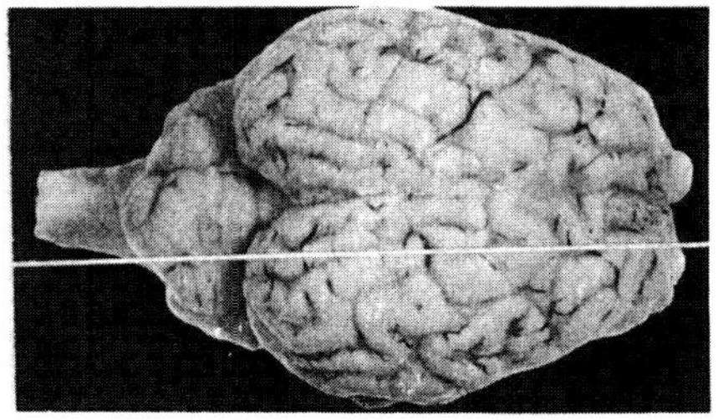

Figure 9

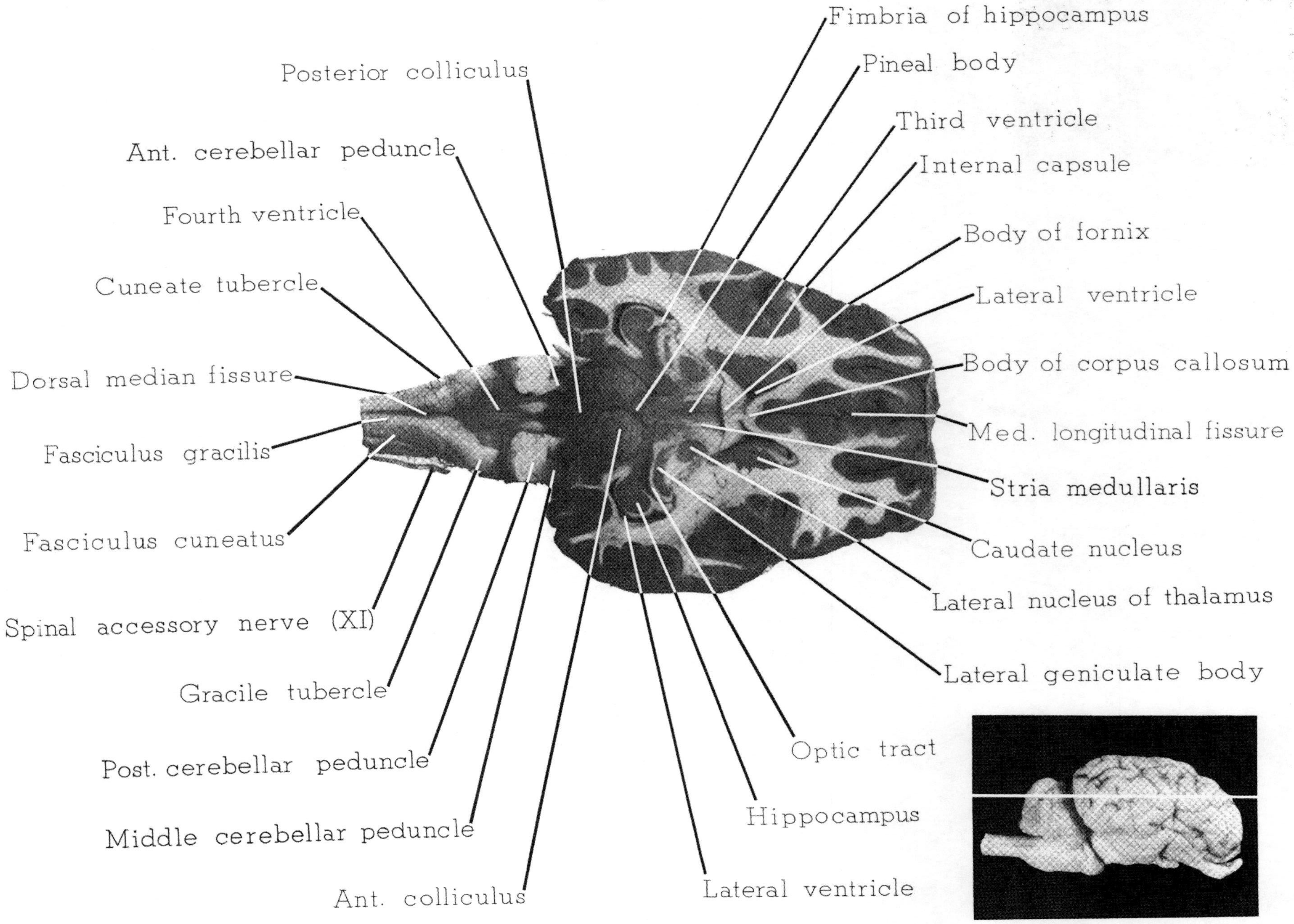

Fimbria of hippocampus
Pineal body
Third ventricle
Internal capsule
Body of fornix
Lateral ventricle
Body of corpus callosum
Med. longitudinal fissure
Stria medullaris
Caudate nucleus
Lateral nucleus of thalamus
Lateral geniculate body
Optic tract
Hippocampus
Lateral ventricle
Posterior colliculus
Ant. cerebellar peduncle
Fourth ventricle
Cuneate tubercle
Dorsal median fissure
Fasciculus gracilis
Fasciculus cuneatus
Spinal accessory nerve (XI)
Gracile tubercle
Post. cerebellar peduncle
Middle cerebellar peduncle
Ant. colliculus

Figure 10

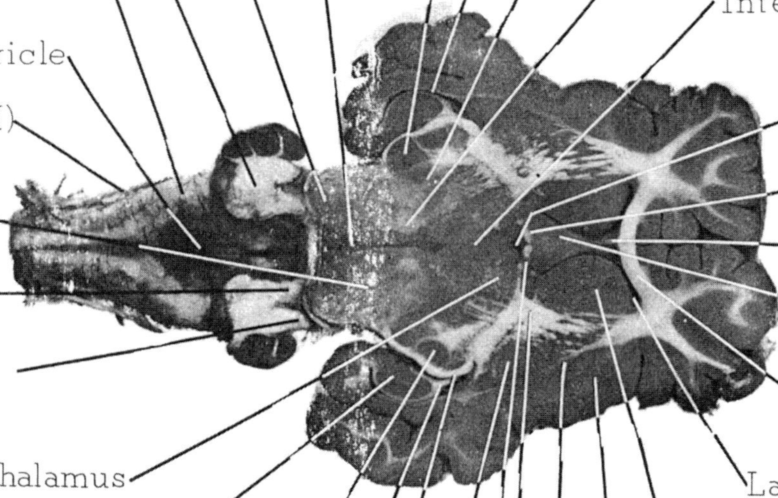

Cerebral aqueduct
Post. colliculus
Post. cerebellar peduncle
Gracile tubercle
Fourth ventricle
Spinal accessory nerve (XI)
Ant. colliculus
Ant. cerebellar peduncle
Middle cerebellar peduncle
Ventral nucleus of thalamus
Hippocampus
Lateral geniculate body
Fimbria of hippocampus
Optic tract
Lentiform nucleus
Pulvinar of thalamus
Lateral ventricle
Post. nucleus of thalamus
Post. commissure
Intermediate mass of thalamus
Third ventricle
Ant. column of fornix
Septum pellucidum
Septal nuclei
Genu of corpus callosum
Lateral ventricle
Caudate nucleus
Cerebral cortex
External capsule
Internal capsule
Foramen of Monro

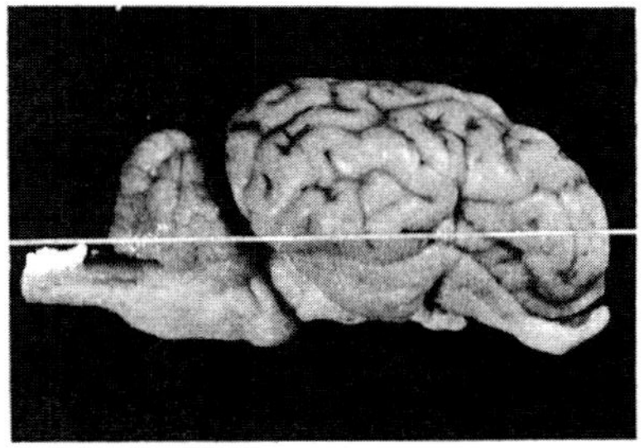

Figure 11

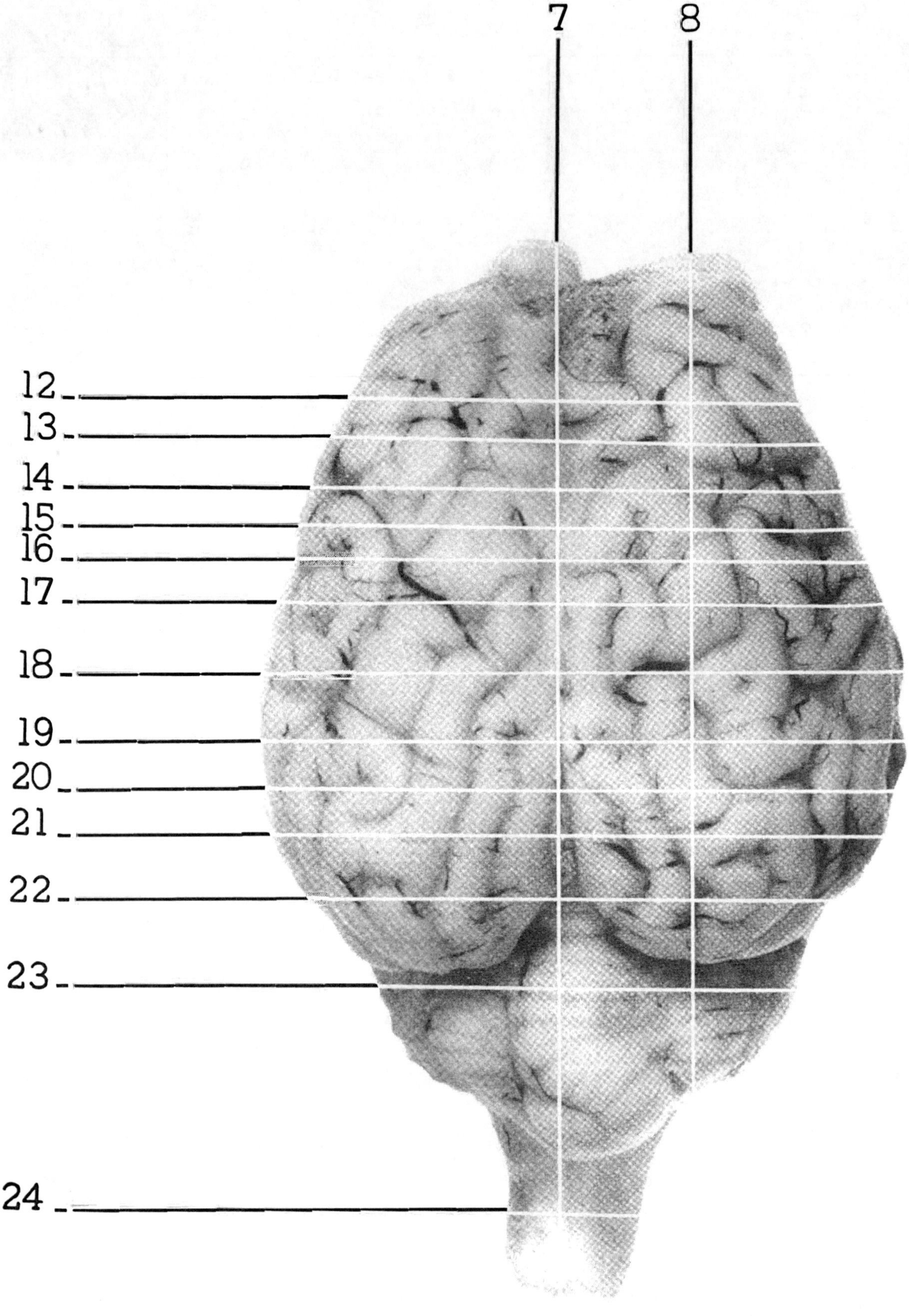

Figure 12

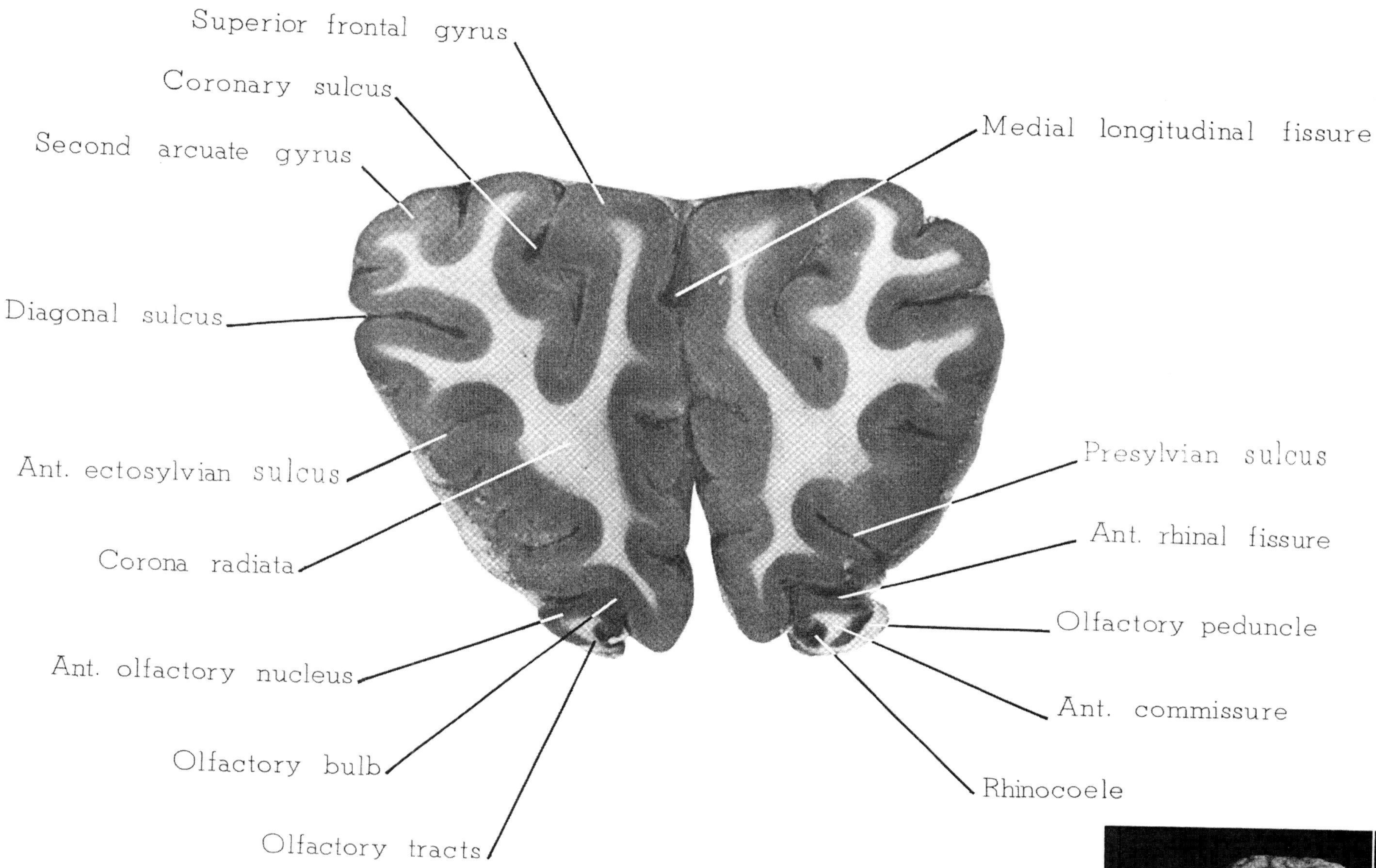
Superior frontal gyrus
Coronary sulcus
Second arcuate gyrus
Diagonal sulcus
Ant. ectosylvian sulcus
Corona radiata
Ant. olfactory nucleus
Olfactory bulb
Olfactory tracts
Medial longitudinal fissure
Presylvian sulcus
Ant. rhinal fissure
Olfactory peduncle
Ant. commissure
Rhinocoele

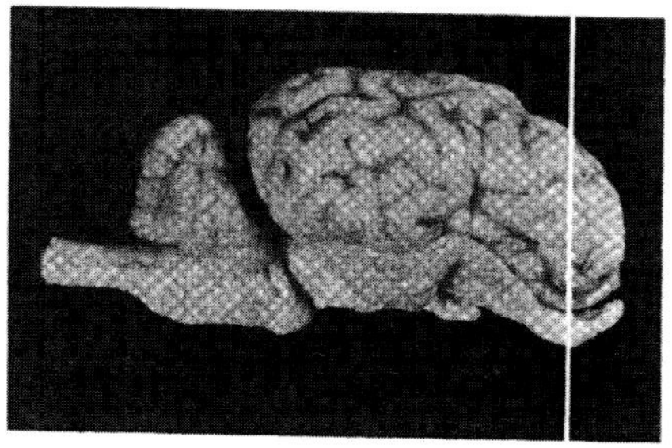

Figure 13

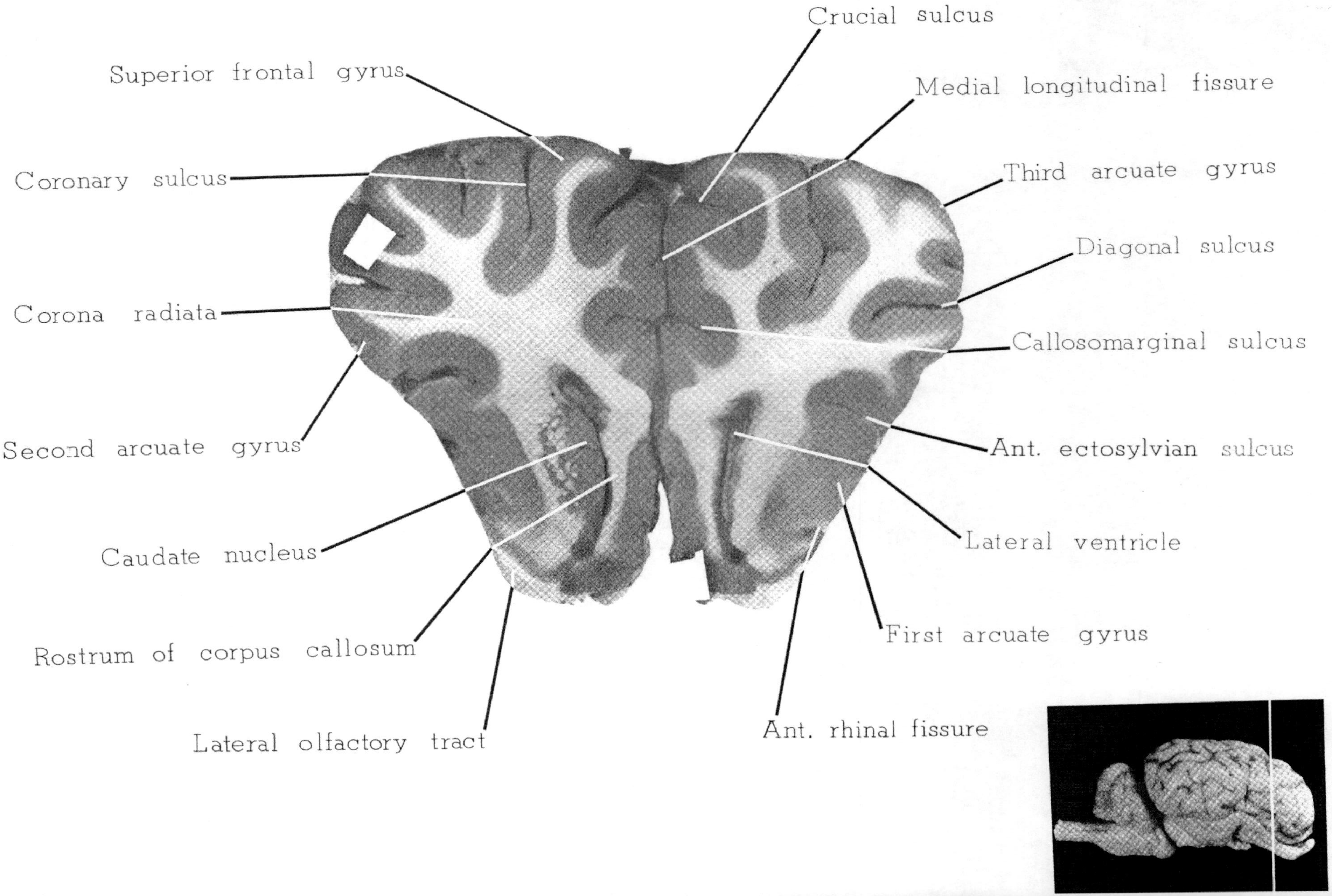

Superior frontal gyrus
Coronary sulcus
Corona radiata
Second arcuate gyrus
Caudate nucleus
Rostrum of corpus callosum
Lateral olfactory tract
Crucial sulcus
Medial longitudinal fissure
Third arcuate gyrus
Diagonal sulcus
Callosomarginal sulcus
Ant. ectosylvian sulcus
Lateral ventricle
First arcuate gyrus
Ant. rhinal fissure

Figure 14

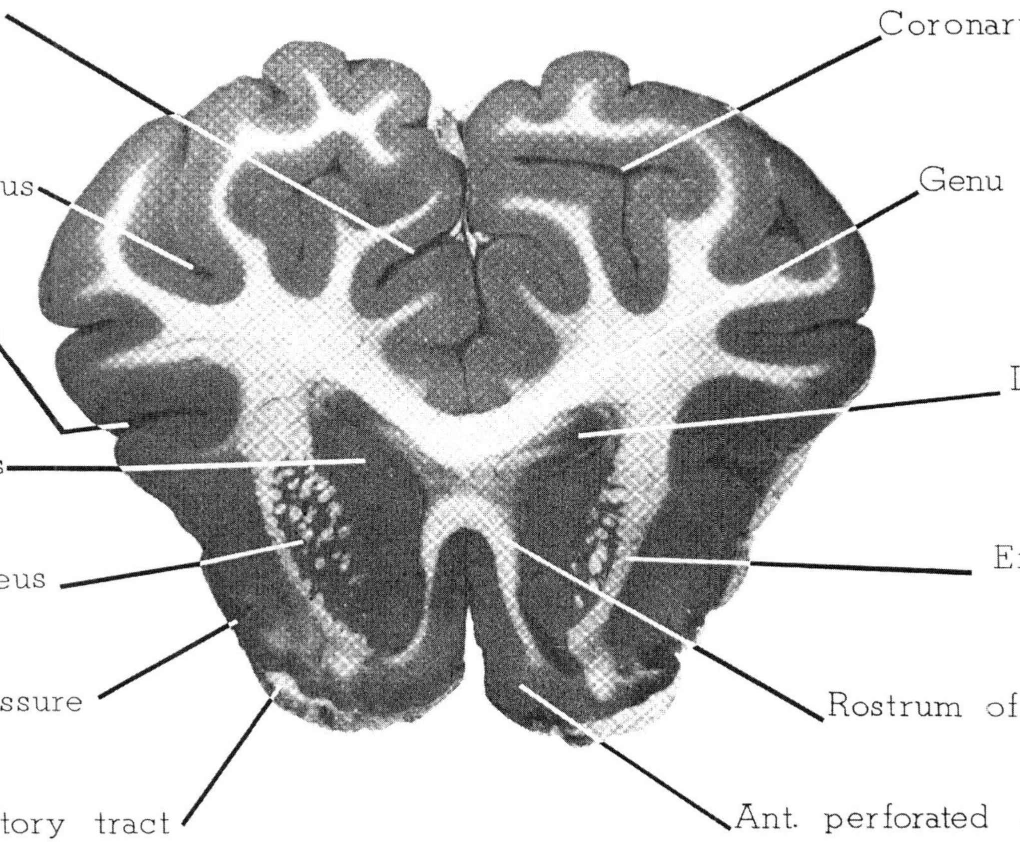
Intercalary sulcus
Ant. suprasylvian sulcus
Ant. ectosylvian sulcus
Caudate nucleus
Lentiform nucleus
Ant. Rhinal fissure
Lateral olfactory tract
Coronary sulcus
Genu of corpus callosum
Lateral ventricle
External capsule
Rostrum of corpus callosum
Ant. perforated substance

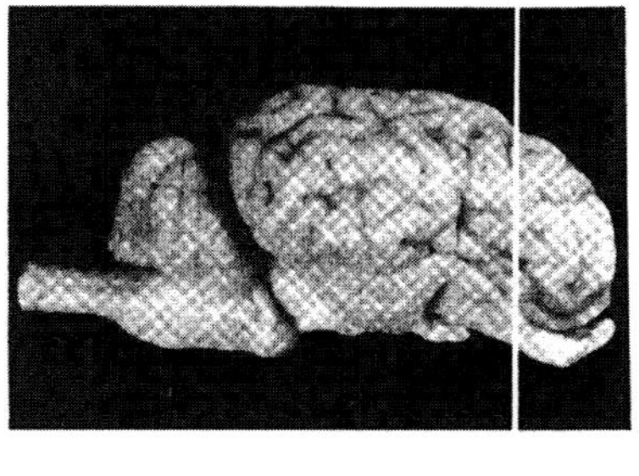

Figure 15

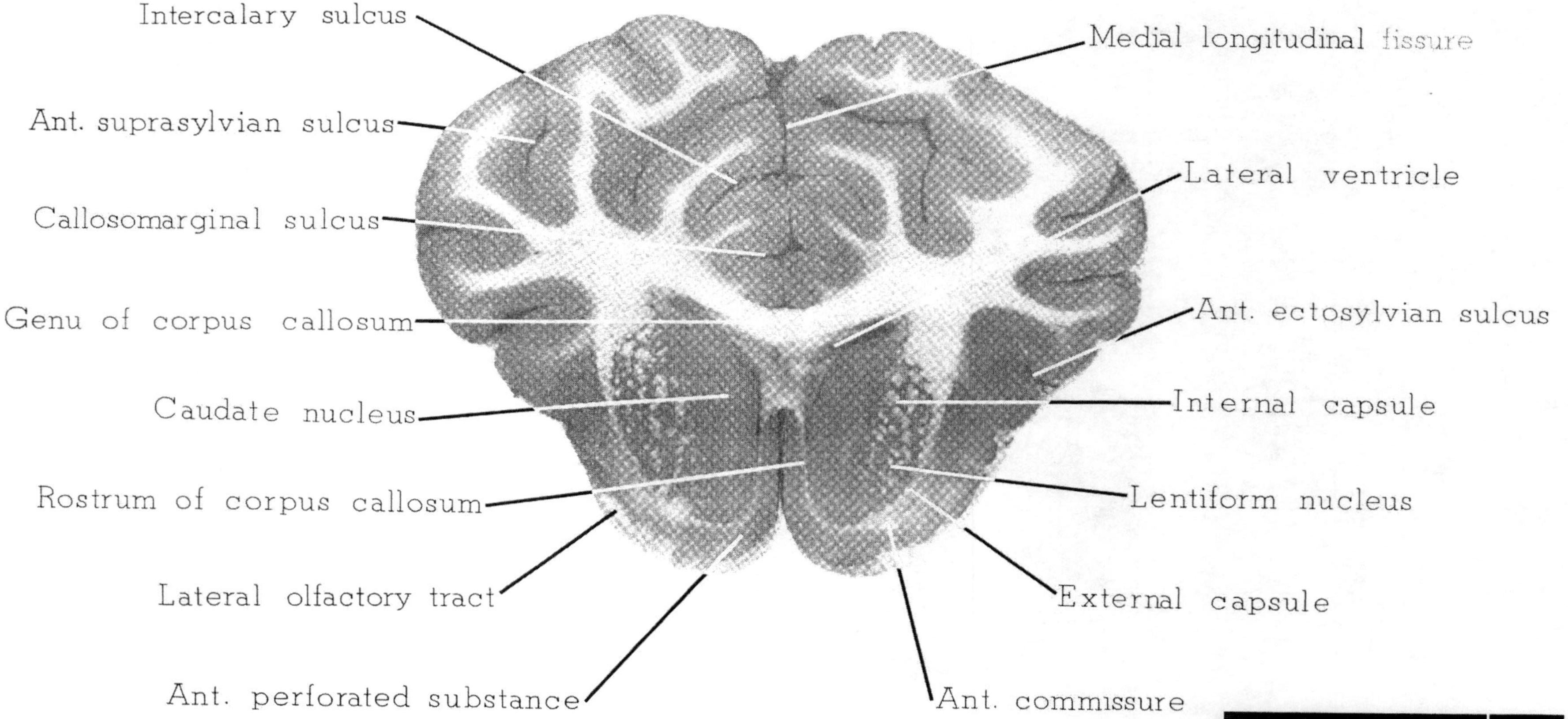

Intercalary sulcus
Ant. suprasylvian sulcus
Callosomarginal sulcus
Genu of corpus callosum
Caudate nucleus
Rostrum of corpus callosum
Lateral olfactory tract
Ant. perforated substance
Medial longitudinal fissure
Lateral ventricle
Ant. ectosylvian sulcus
Internal capsule
Lentiform nucleus
External capsule
Ant. commissure

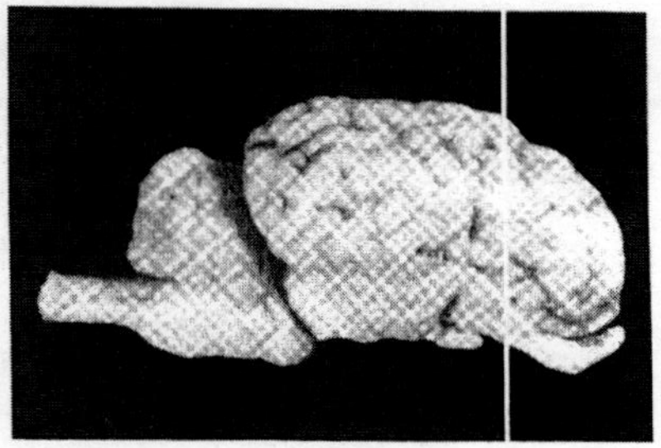

Figure 16

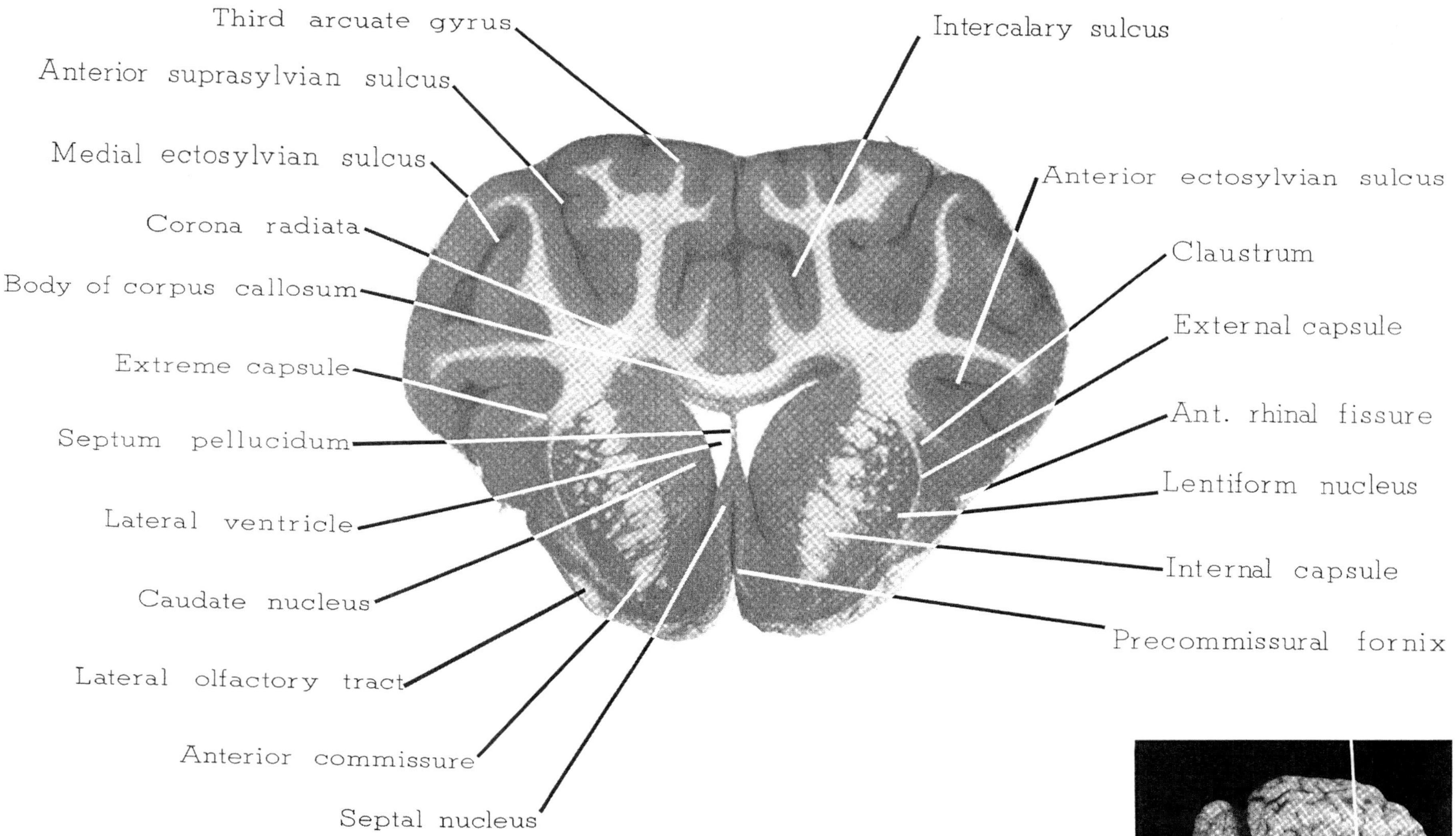
Third arcuate gyrus
Anterior suprasylvian sulcus
Medial ectosylvian sulcus
Corona radiata
Body of corpus callosum
Extreme capsule
Septum pellucidum
Lateral ventricle
Caudate nucleus
Lateral olfactory tract
Anterior commissure
Septal nucleus
Intercalary sulcus
Anterior ectosylvian sulcus
Claustrum
External capsule
Ant. rhinal fissure
Lentiform nucleus
Internal capsule
Precommissural fornix

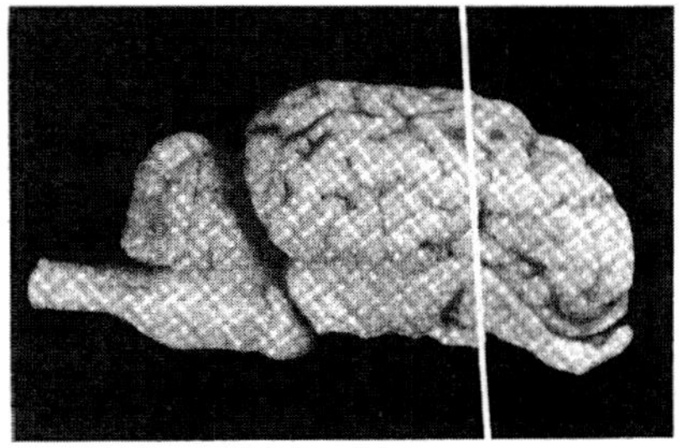

Figure 17

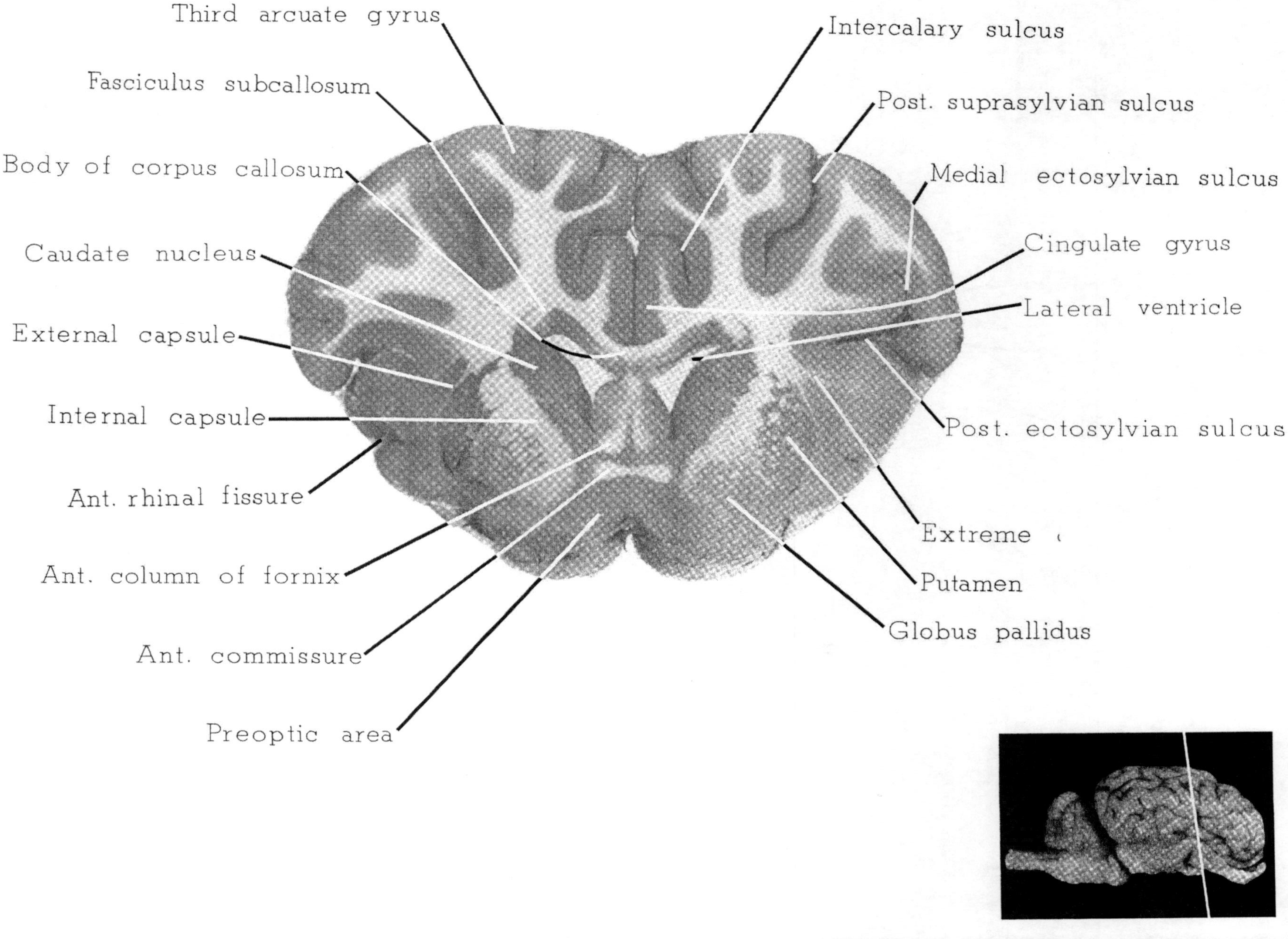
Third arcuate gyrus
Fasciculus subcallosum
Body of corpus callosum
Caudate nucleus
External capsule
Internal capsule
Ant. rhinal fissure
Ant. column of fornix
Ant. commissure
Preoptic area
Intercalary sulcus
Post. suprasylvian sulcus
Medial ectosylvian sulcus
Cingulate gyrus
Lateral ventricle
Post. ectosylvian sulcus
Extreme
Putamen
Globus pallidus

Figure 18

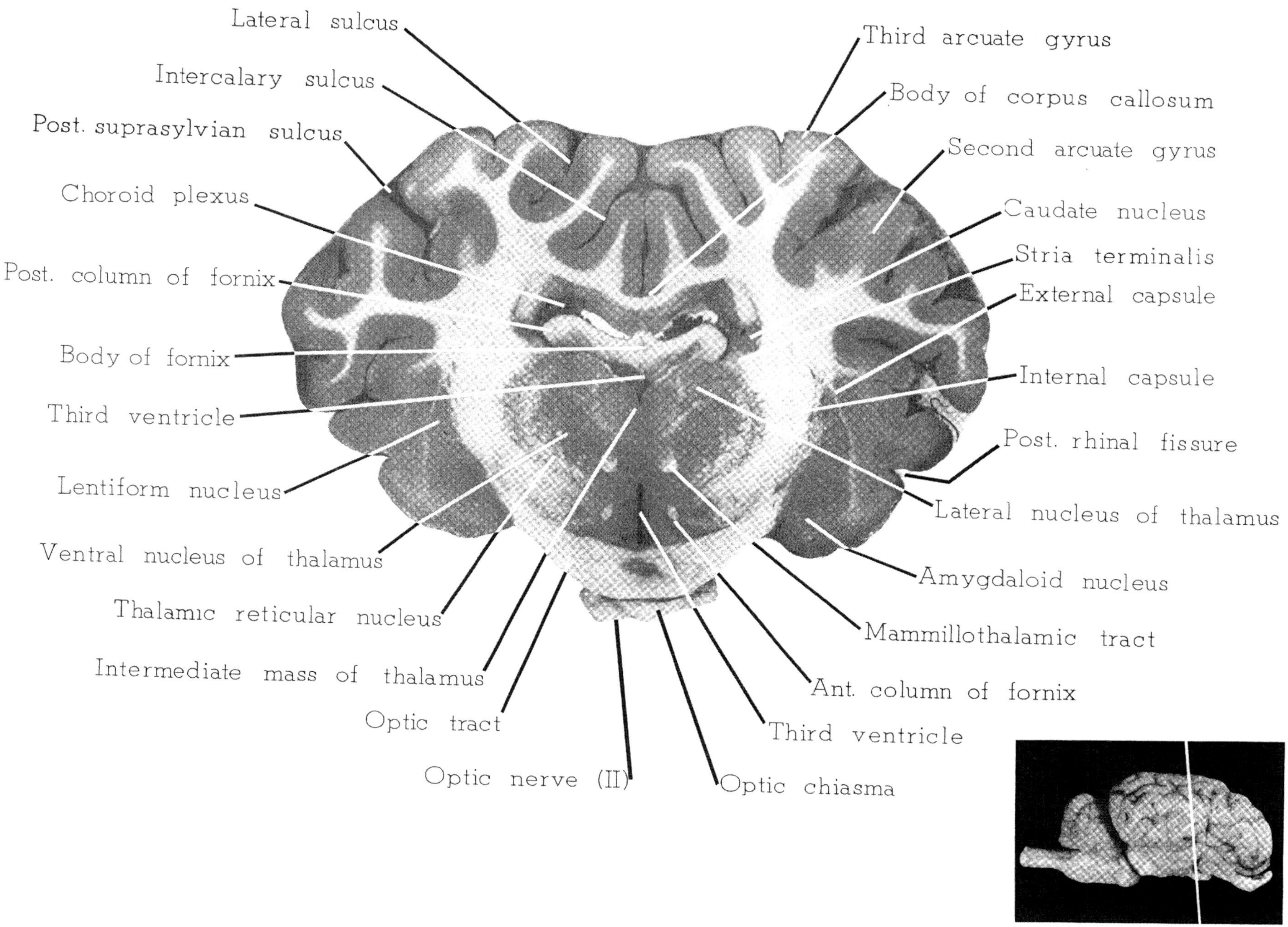
Lateral sulcus
Intercalary sulcus
Post. suprasylvian sulcus
Choroid plexus
Post. column of fornix
Body of fornix
Third ventricle
Lentiform nucleus
Ventral nucleus of thalamus
Thalamic reticular nucleus
Intermediate mass of thalamus
Optic tract
Optic nerve (II)
Optic chiasma
Third ventricle
Ant. column of fornix
Mammillothalamic tract
Amygdaloid nucleus
Lateral nucleus of thalamus
Post. rhinal fissure
Internal capsule
External capsule
Stria terminalis
Caudate nucleus
Second arcuate gyrus
Body of corpus callosum
Third arcuate gyrus

Figure 19

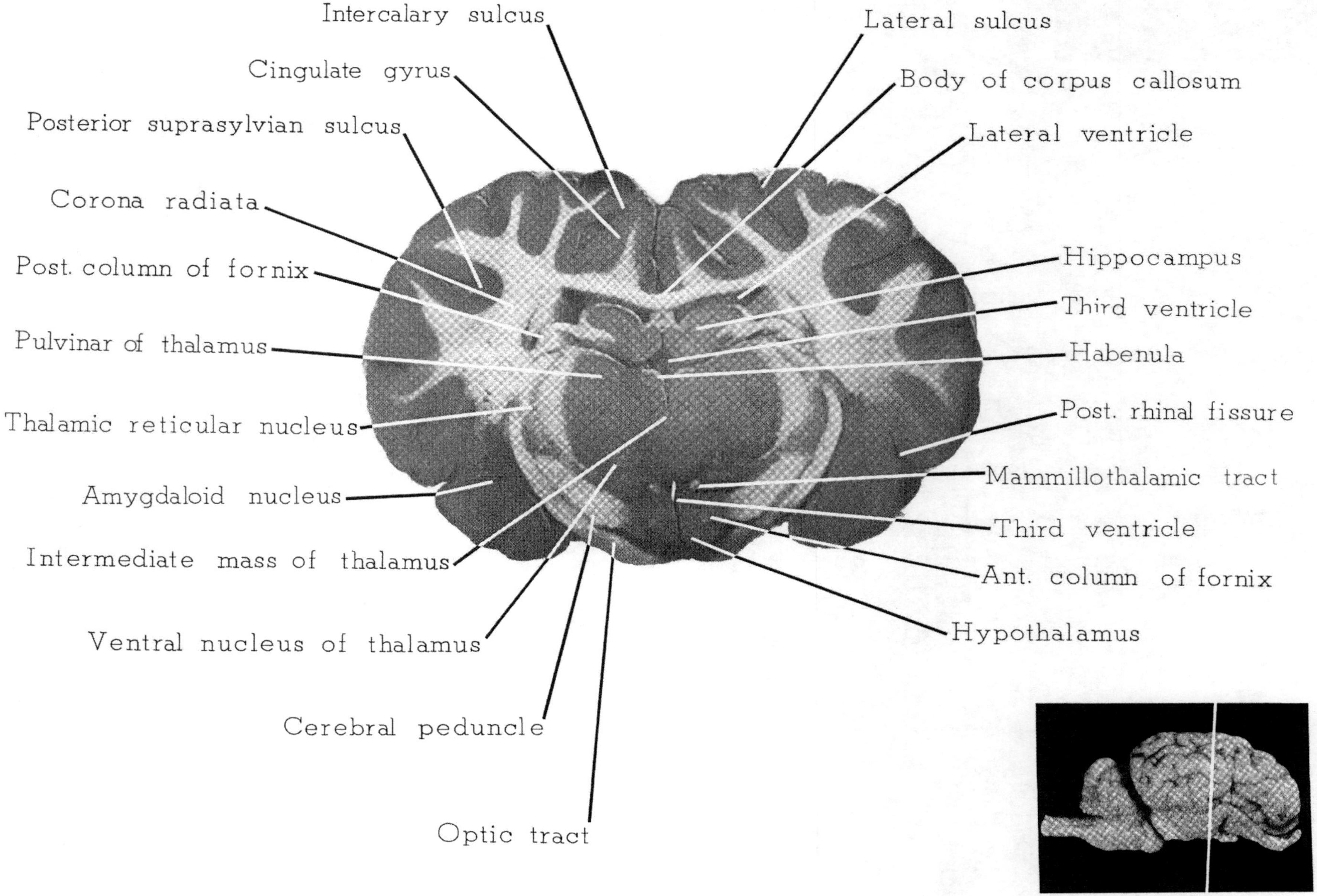

Intercalary sulcus
Cingulate gyrus
Posterior suprasylvian sulcus
Corona radiata
Post. column of fornix
Pulvinar of thalamus
Thalamic reticular nucleus
Amygdaloid nucleus
Intermediate mass of thalamus
Ventral nucleus of thalamus
Cerebral peduncle
Optic tract
Lateral sulcus
Body of corpus callosum
Lateral ventricle
Hippocampus
Third ventricle
Habenula
Post. rhinal fissure
Mammillothalamic tract
Third ventricle
Ant. column of fornix
Hypothalamus

Figure 20

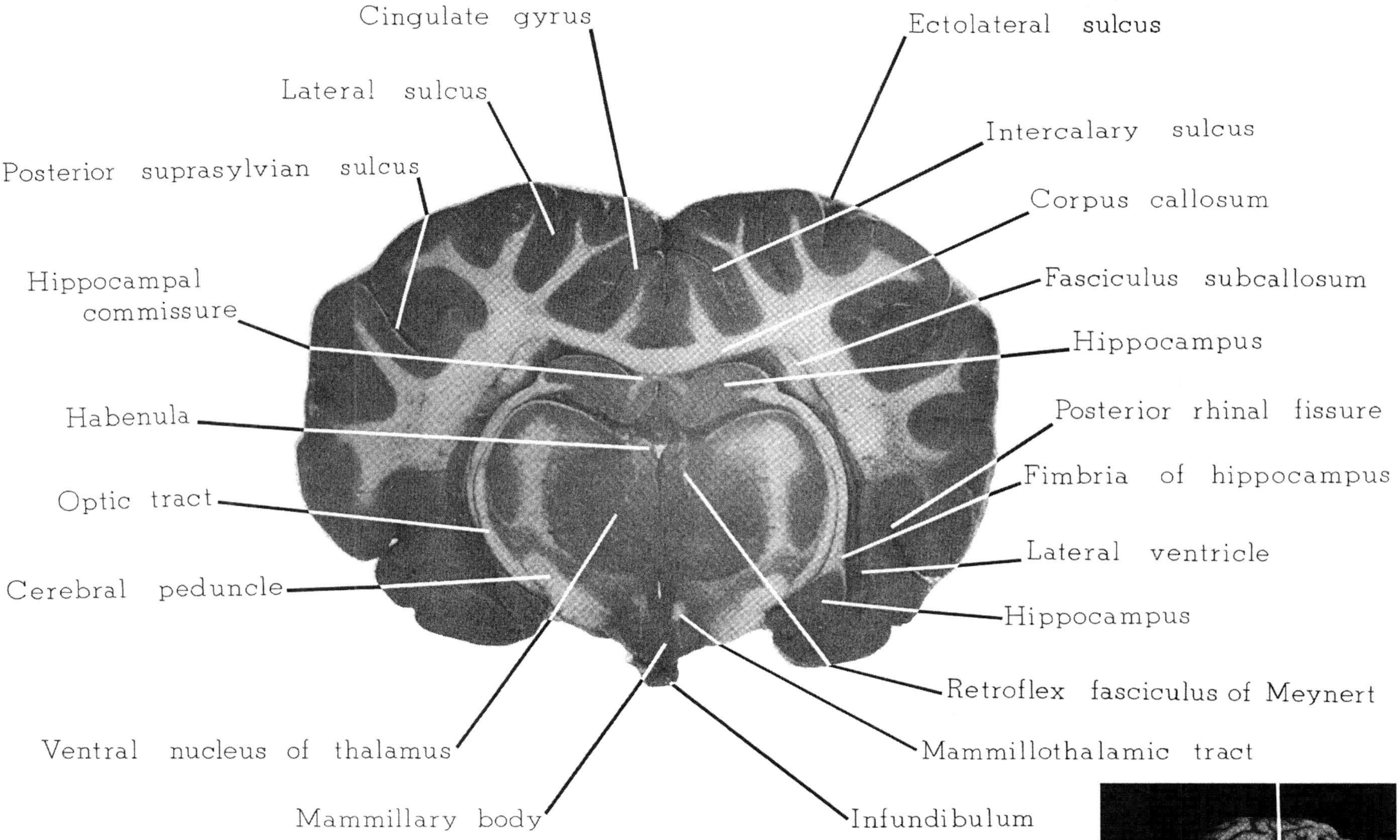
Cingulate gyrus
Ectolateral sulcus
Lateral sulcus
Intercalary sulcus
Posterior suprasylvian sulcus
Corpus callosum
Hippocampal commissure
Fasciculus subcallosum
Hippocampus
Habenula
Posterior rhinal fissure
Fimbria of hippocampus
Optic tract
Lateral ventricle
Cerebral peduncle
Hippocampus
Retroflex fasciculus of Meynert
Ventral nucleus of thalamus
Mammillothalamic tract
Mammillary body
Infundibulum

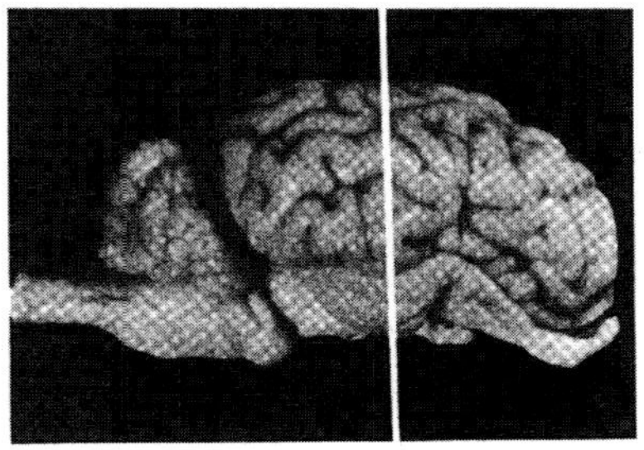

Figure 21

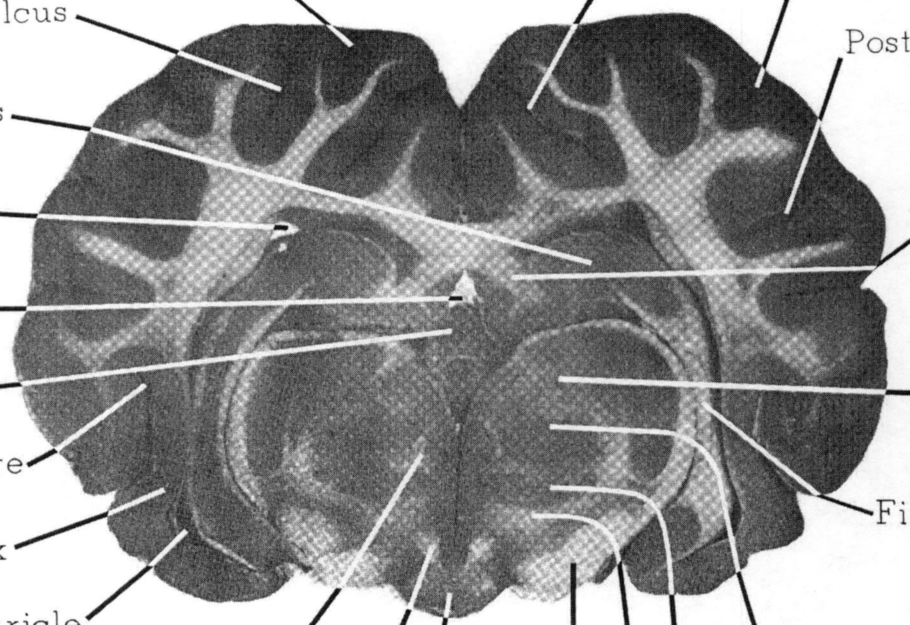

Entolateral sulcus
Lateral sulcus
Hippocampus
Lateral ventricle
Third ventricle
Pineal body
Posterior rhinal fissure
Pyriform cortex
Lateral ventricle
Retroflex fasiculus of Meynert
Mammillothalamic tract
Mammillary body
Intercalary sulcus
Ectolateral sulcus
Posterior suprasylvian fissure
Hippocampal commissure
Pulvinar of the thalamus
Fimbria of the hippocampus
Medial nucleus of the thalamus
Ventral nucleus of the thalamus
Lemniscal fibers
Cerebral peduncle

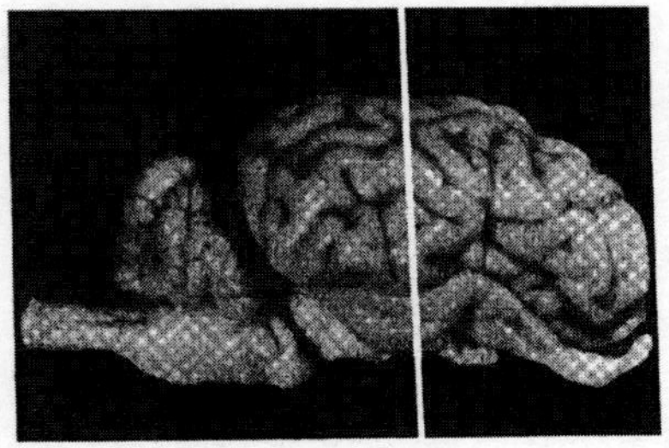

Figure 22

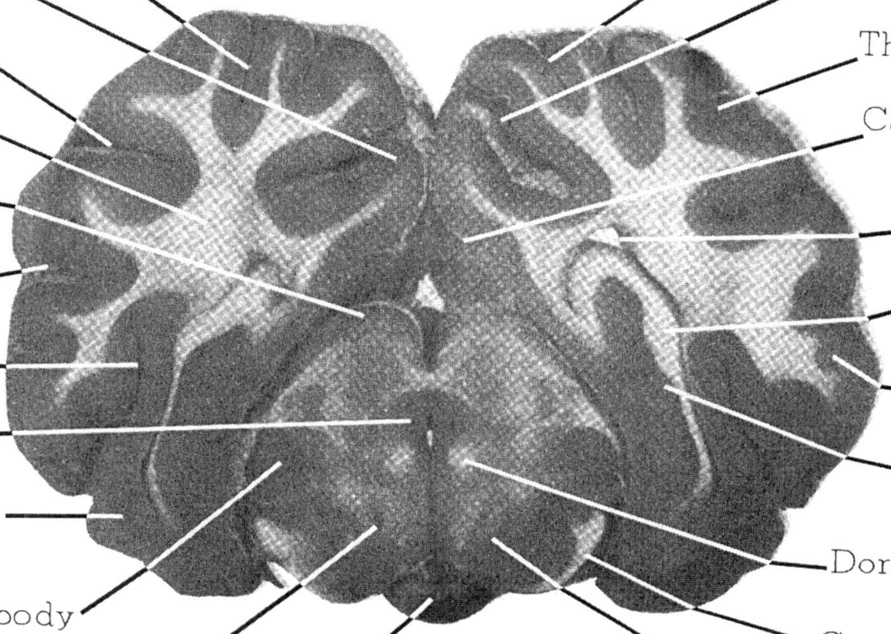
Lateral sulcus
Cingulate gyrus
Ectolateral sulcus
Corona radiata
Ant. colliculus
Post. suprasylvian sulcus
Post. rhinal fissure
Central gray substance
Pyriform cortex
Medial geniculate body
Nucleus ruber
Mammillary body
Entolateral sulcus
Intercalary sulcus
Third arcuate gyrus
Callosomarginal sulcus
Lateral ventricle
Fimbria of hippocampus
Second arcuate gyrus
Hippocampus
Dorsal longitudinal fasciculus
Cerebral peduncle
Substantia nigra

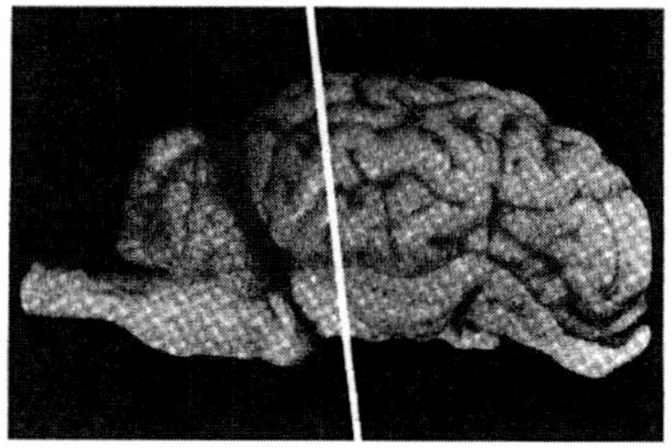

Figure 2 3

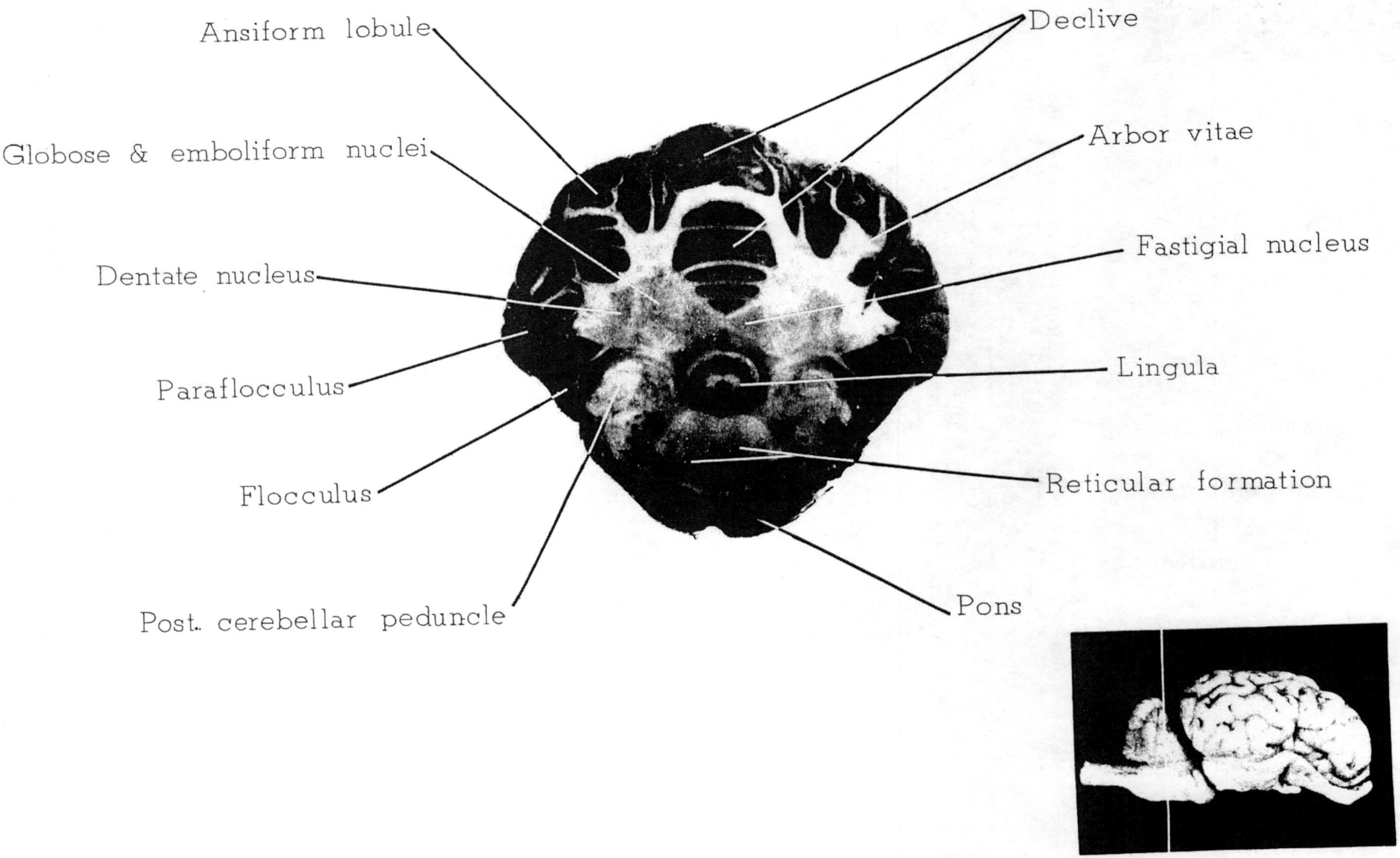
Ansiform lobule
Globose & emboliform nuclei
Dentate nucleus
Paraflocculus
Flocculus
Post. cerebellar peduncle
Declive
Arbor vitae
Fastigial nucleus
Lingula
Reticular formation
Pons

Figure 24

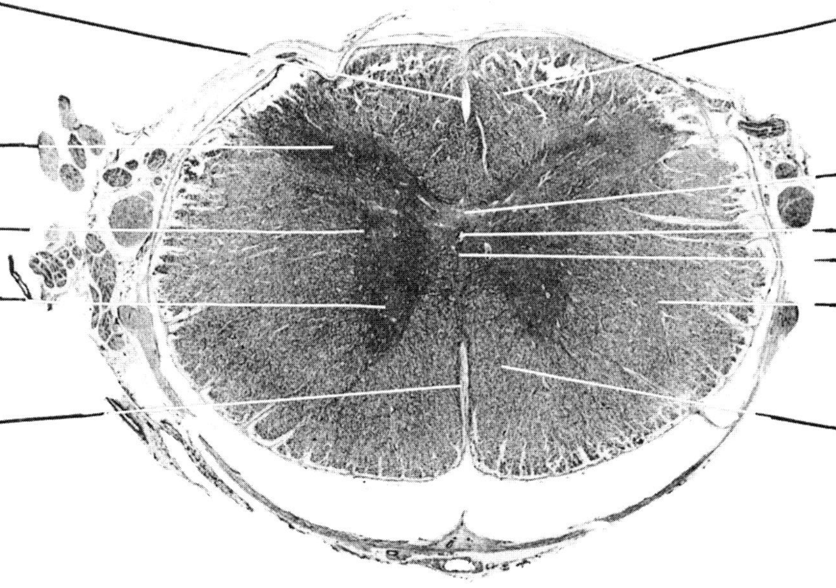

Dorsal median fissure
Dorsal gray column
Lateral gray column
Ventral gray column
Ventral median fissure
Dorsal funiculus
Dorsal gray commissure
Central canal
Ventral gray commissure
Lateral funiculus
Ventral funiculus

V. GLOSSARY

A

<u>Abducent nerve</u> - cranial nerve number 6, somatic motor nerve to external rectus muscle.

<u>Amygdaloid nucleus</u> - subcortical nuclei of the limbic system involved in behavioral functions.

<u>Ansiform lobule</u> - a division of the neocerebellum concerned with timing and guiding of voluntary movements.

<u>Anterior cerebellar peduncle</u> (Brachium conjunctivum) - major efferent pathway from the neocerebellum to cerebral cortex and lower brain stem centers.

<u>Anterior colliculus</u> - anterior nucleus of the lamina quadrigemina, or tectum, mediating optic reflexes.

<u>Anterior column of fornix</u> - that portion of the fornix anterior to the body of the fornix which passes to the maximmary bodies and septal nuclei serving primarily as a motor pathway from the hippocampus to those centers.

<u>Anterior commissure</u> - a tract interconnecting the amygdaloid nuclei and olfactory bulbs of both sides of the forebrain.

<u>Anterior ectosylvian sulcus</u> - a fold or groove separating the first and second arcuate gyri of the cerebral cortex.

<u>Anterior olfactory nucleus</u> - a secondary relay center located in the olfactory peduncles for olfactory sensations.

<u>Anterior parolfactory sulcus</u> - a fold separating the frontal cerebral cortex from the parolfactory gyrus and medial olfactory tract.

<u>Anterior perforated substance</u> (olfactory tubercule) - a secondary olfactory center receiving fibers of the lateral and medial olfactory tracts and discharging to higher limbic centers.

<u>Anterior rhinal fissure</u> - a groove separating the pyriform lobe (allocortex) from the cerebral cortex (isocortex).

<u>Anterior suprasylvian sulcus</u> - a fold separating the second and third arcuate gyri of the cerebral cortex.

<u>Aqueduct of Sylvius</u> - see cerebral aqueduct.

<u>Arbor vitae</u> - the fibrous center of the cerebellum composed of afferent and efferent pathways to and from the cerebellum.

<u>Auditory nerve</u> - cranial nerve number 8, special sensory nerve from the inner ear.

B

Band of Broca - a ridge created by fibers passing between pyriform lobe and medial surface of the cerebral hemisphere and located anterior to the optic tracts.

Body of corpus callosum - a major division of the corpus callosum. A tract, or commissure, connecting the left and right cerebral cortices.

Body of fornix - the medial division which gives rise to the anterior column of the fornix and which is created by fusion of the posterior columns of the fornix.

Brachium conjunctivum - see anterior cerebellar peduncle.

Brachium pontis - see middle cerebellar peduncle.

C

Callosomarginal sulcus - see cingulate sulcus.

Caudate nucleus - a center of the neostriatum involved with other extrapyramidal nuclei in locomotive patterns.

Central canal - a central space communicating with the fourth ventricle and located within the spinal cord.

Central gray substance - a group of neurons surrounding the cerebral aqueduct and functioning as relay nuclei for hypothalamic discharge to spinal centers.

Central lobule - a division of the paleocerebellum concerned with regulation of reflex tonus.

Cerebellar cortex - all external gray matter of the cerebellum.

Cerebellar hemisphere - lateral division of the cerebellum.

Cerebral aqueduct (Aqueduct of Sylvius) - a narrow canal connecting the third and fourth ventricles and located ventrally to the lamina quadrigemina.

Cerebral cortex - all external gray matter dorsal to the rhinal fissure of the telencephalon.

Cerebral peduncles - rope-like bundles which are the motor pathways of the telencephalon and located in the base of the midbrain.

Choroid plexus - tuft-like projection of neural epithelial tissue and capillary bed into the ventricles.

Cingulate gyrus (fornicate gyrus) - a division of the juxtallocortex located dorsal to the corpus callosum.

Cingulate sulcus (callosomarginal sulcus) - a groove in the cerebral cortex separating the cingulate gyrus from neocortical gyri.

Clava - see gracile tubercle.

Claustrum - a division of the neocortex lateral to the lentiform nucleus.

Corona radiata - a fan-like pattern of fibers radiating from the internal capsule which pass to and from the cerebral cortex.

Coronary sulcus - a fold in the cerebral cortex separating the third arcuate gyrus from the superior frontal gyrus.

Corpora quadrigemina - see lamina quadrigemina.

Corpus callosum - a major commissure connecting the left and right cerebral hemispheres.

Crucial sulcus - a fold in the cerebral cortex separating the superior frontal gyrus from the medial cerebral gyri.

Crus of fornix - see posterior column of fornix.

Culmen - a division of the paleocerebellum involved in regulation of reflex tonus.

Cuneate tubercle - a relay nucleus located on the dorsolateral surface of the medulla oblongata for general somatic sensory sensations to the thalamus and cerebral cortex.

D

Declive - a division of the paleocerebellum involved in regulation of reflex tonus.

Dentate nucleus - major efferent center from neocerebellum to ruber nucleus and thalamus.

Diagonal sulcus - fold in cerebral cortex located between anterior ectosylvian and suprasylvian sulci.

Dorsal funiculus - dorsal white matter composed of the fasciculi cuneatus and gracilis of the spinal cord.

Dorsal gray column - location of many second order sensory neurons of the spinal nerves.

Dorsal gray commissure - medial gray matter dorsal to central canal of the spinal cord.

Dorsal longitudinal fasciculus - a hypothalamic motor pathway located ventrally to the central gray substance.

Dorsal median fissure - dorsal groove partially separating spinal cord into left and right halves.

E

Ectolateral sulcus - a fold in the cerebral cortex which separates the suprasylvian gyrus from the lateral gyrus.

Emboliform nucleus - a motor center from the paleocerebellum to lower brain stem centers.

Endorhinal fissure - a fold in the floor of the telencephalon which separates the anterior perforated substance and the lateral olfactory tract.

Entolateral sulcus - most dorsomedial groove on the cerebral cortex.

External capsule - separates putamen and globus pallidus from claustrum and consists of fibers to and from the cerebral cortex and lentiform nuclei.

Extreme capsule - partially separates claustrum from cerebral cortex and consists of fibers to and from these regions.

F

Facial nerve - cranial nerve number 7 involved in sensory and motor functions of facial musculature and gustatory sensations.

Fasciculus cuneatus - lateral division of the dorsal funiculus of the spinal cord.

Fasciculus gracilis - medial division of the dorsal funiculus of the spinal cord.

Fasciculus subcallosum - associative cortical tract which also projects to caudate nucleus and is ventrally located to the corpus callosum.

Fastigial nucleus - efferent relay nucleus from archicerebellum to lower brain stem centers.

Fimbria of hippocampus - rope-like tract originating from hippocampus and forming posterior column of fornix.

First arcuate gyrus - a convolution, or ridge, dorsal to rhinal fissure.

Flocculus - a division of the archicerebellum involved in regulation of vestibular reflexes.

Foramen of Monro - opening between lateral and third ventricles for passage of cerebrospinal fluid.

Fornix - major efferent pathway from hippocampus to lower brain stem centers.

Fourth ventricle - a cavity found ventrally to the cerebellum.

G

Genu of corpus callosum - a division of the corpus callosum at the point of its anterio-ventrally directed flexion.

Globose nucleus - an efferent relay nucleus of the paleocerebellum.

Globus pallidus - paleostriatal nucleus, or ventral component of the lentiform nucleus, involved in locomotive patterns.

Glossopharyngeal nerve - cranial nerve number 9 involved in visceral sensory and motor functions.

Gracile tubercle (Clava) - a relay nucleus located on the dorsomedial surface of the medulla oblongata involved in general somatic sensations to the thalamus and cerebral cortex.

H

Habenula - an epithalamic nucleus involved in the functions of the limbic system.

Habenulopeduncular tract - see retroflex fasciculus of Meynert.

Hippocampal commissure - a fiber tract interconnecting the left and right hippocampal complexies.

Hippocampus (Ammon's horn) - a major visceral-somatic coordination center of the limbic system.

Hypothalamus - concerned with visceral and behavioral functions and located on the floor of the diencephalon.

I

Infundibulum - a tube-like structure connecting the hypothalamus to the pituitary gland.

Intercalary sulcus - a fold in the cerebral cortex dorsal to the cingulate gyrus.

Intermediate mass of thalamus - a division of the thalamus crossing the third ventricle.

Internal capsule - a bundle of fibers passing through the basal nuclei on their way to and from the cerebral cortex, basal nuclei, and lower brain stem centers.

L

Lamina quadrigemina - the roof of the mesencephalon, or midbrain.

Lamina terminalis - the anterior end of the forebrain.

Lateral funiculus - laterally placed ascending and descending fiber pathways found in the spinal cord.

Lateral geniculate body - optic relay center to the visual area of the cerebral cortex (occipital region).

Lateral gray column - visceral motor nuclei located ventrolaterally to the dorsal gray column of the spinal cord.

Lateral nucleus of thalamus - general somatic sensory relay nuclei to the cerebral cortex.

Lateral olfactory gyrus (prepyriform cortex) - a division of the allocortex of the limbic system.

Lateral olfactory tract - a group of fibers carrying olfactory impulses to the amygdaloid nuclei and the allocortices.

Lateral sulcus - a fold in the cerebral cortex separating the ectolateral gyrus from the lateral gyrus.

Lateral ventricle - cavity found within each cerebral hemisphere.

Lemniscal fibers - major somatic sensory and auditory projection systems to forebrain.

Lentiform nucleus - nuclear complex formed by putamen, nucleus of the neostriatum, and globus pallidus of the paleostriatum.

Lingula - a division of the paleocerebellum involved in regulation of postural tonus.

M

Mammillary body - involved in limbic functions and located on the floor of the diencephalon posterior to the hypothalamus.

Mammillothalamic tract - an efferent pathway from the mammillary body to the thalamus.

Medial ectosylvian sulcus - a fold in the cerebral cortex dorsal to the pseudosylvian sulcus.

Medial geniculate body - auditory relay nucleus to auditory region of cerebral cortex (temporal lobe).

Medial longitudinal fissure - dorsal groove which partially divides the telencephalon into left and right hemispheres.

Medial nucleus of thalamus - thalamic nuclei located medially to the third ventricle. They project to the hypothalamus, prefrontal lobe of cerebrum, and subcortical nuclei.

Medial olfactory gyrus - a convolution bordered anterio-medially by the ant. parolfactory sulcus and posterio-laterally by the post. parolfactory sulcus.

Medial olfactory tract - a pathway carrying olfactory impulses to the olfactory tubercule and other medial basal forebrain centers.

Medulla oblongata - most posterior division of brain stem involved in visceral reflexes and major relay centers for cranial nerves.

Middle cerebellar peduncle (Brachium pontis) - major afferent pathway to neo-cerebellum from the cerebral cortex.

N

Nodule - medial division of the archicerebellum.

O

Oculomotor nerve - cranial nerve number 3, somatic and visceral motor functions.

Olfactory bulb - relay center for second order neurons from nasal mucosa.

Olfactory peduncle - name given to olfactory tracts as they leave the olfactory bulb.

Olfactory tracts - pathways which carry olfactory impulses from olfactory bulbs to higher forebrain centers.

Olive - a group of neurons located in the ventrolateral walls of the medulla oblongata and involved in auditory and cerebellar functions.

Optic chiasma - located ventrally to diencephalon and involved in partial decussation of optic nerves.

Optic nerve - cranial nerve number 2, carries optic impulses from eye.

Optic tract - that portion of the optic system posterior to the optic chiasma and projecting to the anterior colliculus, hypothalamus, and lateral geniculate body.

P

Paraflocculus - one of the two lateral divisions of the archicerebellum.

Paramedian lobule - a division of the neocerebellum involved in timing of voluntary motor actions.

Pineal body - a division of the epithalamus and involved in endocrine functions.

Pons - a large ventral ridge marking the anterior limit of the medulla oblongata and composed of the pontine nuclei and middle cerebellar peduncles.

Pontine nuclei - relay nuclei for pathways from the cerebral cortex to the neocerebellum.

Posterior cerebellar peduncle (restiform body) - major afferent pathway from spinal cord to paleocerebellum.

Posterior colliculus - auditory reflex center located in roof of midbrain.

Posterior column of fornix - anteriorly directed continuation of the fimbria of the hippocampus.

Posterior commissure - a pathway interconnecting various tectal, or mesencephalic, nuclei with diencephalic centers.

Posterior ectosylvian sulcus - a groove in the posterior cerebral cortex separating the first and second arcuate gyri.

Posterior nucleus of thalamus - thalamic nuclei posterior to the lateral and ventral thalamic nuclei and involved in projection to associative cerebral cortex.

Posterior parolfactory sulcus - a groove which separates medial olfactory gyrus from anterior perforated substance.

Posterior rhinal fissure - continuation of anterior rhinal fissure beyond pseudosylvian sulcus.

Posterior syprasylvian sulcus - a groove in the cerebral cortex separating the second and third arcuate gyrus.

Precommissural fornix - a division of the anterior column of the fornix passing beyond the anterior commissure to end in rostal basal centers.

Primary fissure - a fold in the cerebellum separating the culmen from the declive.

Preoptic area - the most rostal extent of the diencephalon.

Presylvian sulcus - a fold in the cerebral cortex anterior to the first arcuate gyrus.

Pesudosylvian sulcus - a vertical fold in the first arcuate gyrus.

Pulvinar of thalamus - a posterior thalamic associative nucleus which projects to the parietal lobe of the cerebral cortex.

Putamen - lateral division of the neostriatum. Forms dorsolateral component of lentiform complex.

Pyramid - a division of the paleocerebellum.

Pyramidal tract (corticospinal tract) - a heavily myelinated tract running along the floor of the brain stem to cranial motor nuclei and motor nuclei of spinal cord.

Pyriform lobe - located ventrally to the rhinal fissure. A division of the allocortex involved in olfactory functions.

R

Restiform body - see posterior cerebellar peduncle.

Reticular formation - a neuropil centrally located in the brain stem and involved in facilitation of the cerebral cortex and lower motor centers of the brain stem.

Retroflex fasciculus of Meynert (habenulopeduncular tract) - efferent pathway from habenula to lower midbrain motor centers.

Rhinocoele - anterior continuation of lateral ventricle into olfactory bulb.

Rostrum of corpus callosum - most rostal extent of the corpus callosum.

Ruber nucleus - extrapyramidal motor nucleus located in the floor of the midbrain and involved in locomotive activity.

S

Second arcuate gyrus - a cerebral convolution dorsal to the ectosylvian sulcus and ventral to the suprasylvian sulcus.

Septal nuclei - basal telencephalic nuclei involved in functions of the limbic system.

Septum pellucidum - the medial telencephalic wall dorsal to the septal nuclei and medially located to the lateral ventricles.

Simple lobule (declive) - a division of the paleocerebellum.

Spinal accessory nerve - cranial nerve number 11 involved in special visceral motor functions.

Spinal cord - continuation of the brain beyond foramen magnum of skull.

Splenium of corpus callosum - most caudal division of the corpus callosum.

Stria medullaris - a pathway from the thalamus to the habenula.

Stria terminalis - a pathway from the amygdaloid nuclei to lower brain stem centers.

Substantia nigra - a subthalamic nucleus involved in the extrapyramidal locomotive system.

Superior frontal gyrus - the frontal motor gyrus in lower mammals.

Syprasylvian sulcus - a cerebral groove which separates the ectolateral gyrus from the third arcuate gyrus.

T

Thalamic reticular nucleus - most rostal continuation of the reticular formation.

Third arcuate gyrus - cerebral convolution located between suprasylvian sulcus and ectolateral sulcus.

Third ventricle - cavity found in the diencephalon.

Trapezoid body - located posteriorly to the pons and involved in auditory functions.

Trigeminal nerve - cranial nerve number 5 involved in general somatic sensory and special visceral motor functions.

Trochlear nerve - cranial nerve number 4 involved in somatic motor function of the superior oblique muscle.

Tuber vermis - a division of the paleocerebellum.

U

Uvula - a division of the paleocerebellum.

V

Vagus nerve - cranial nerve number 10 involved in general and special visceral sensory and motor functions.

Ventral funiculus - ventrally located white matter consisting of descending pathways located in the spinal cord.

Ventral gray column - location of somatic motor neurons of the spinal nerves.

Ventral gray commissure - medial gray matter ventral to central canal of the spinal cord.

Ventral median fissure - ventral groove partially separating spinal cord into left and right halves.

Ventral nucleus of thalamus - thalamic nuclei important as relay nuclei to cerebral cortex for somatic sensory, rubrothalamic, and dentatothalamic pathways.

Vermis - medial division of the cerebellum.

VI. INDEX

A

B

C

D

E

F

G

H

I

J

L

M

N

O

P

R

S

T

U

V